Springer Series in
OPTICAL SCIENCES 109

founded by H.K.V. Lotsch

Springer Series in
OPTICAL SCIENCES

The Springer Series in Optical Sciences, under the leadership of Editor-in-Chief *William T. Rhodes*, Georgia Institute of Technology, USA, provides an expanding selection of research monographs in all major areas of optics: lasers and quantum optics, ultrafast phenomena, optical spectroscopy techniques, optoelectronics, quantum information, information optics, applied laser technology, industrial applications, and other topics of contemporary interest.

With this broad coverage of topics, the series is of use to all research scientists and engineers who need up-to-date reference books.

The editors encourage prospective authors to correspond with them in advance of submitting a manuscript. Submission of manuscripts should be made to the Editor-in-Chief or one of the Editors. See also http://springeronline.com/series/624

Motoichi Ohtsu (Ed.)

Progress in Nano-Electro-Optics IV

Characterization
of Nano-Optical Materials
and Optical Near-Field Interactions

With 123 Figures

 Springer

Professor Dr. Motoichi Ohtsu
Department of Electronics Engineering
School of Engineering
The University of Tokyo
7-3-1 Hongo, Bunkyo-ku, Tokyo 113-8656, Japan
E-mail: ohtsu@ee.t.u-tokyo.ac.jp

ISSN 0342-4111

ISBN 3-540-23236-2 Springer Berlin Heidelberg New York

Library of Congress Cataloging-in-Publication Data

Progress in nano-electro-optics IV : characterization of nano-optical materials and optical near-field interactions / Motoichi Ohtsu (ed.). p.cm. – (Springer series in optical sciences ; v. 109)
Includes bibliographical references and index.
ISBN 3-540-23236-2 (alk. paper)
1. Electrooptics. 2. Nanotechnology. 3. Near-field microscopy. I. Ohtsu, Motoichi. II. Series.
TA1750 .P75 2002   621.381'045–dc21   2002030321

Springer is a part of Springer Science+Business Media.

springeronline.com

© Springer-Verlag Berlin Heidelberg 2005
Printed in The Netherlands

Typesetting and prodcution: PTP-Berlin, Protago-TEX-Production GmbH, Berlin
Cover concept by eStudio Calamar Steinen using a background picture from The Optics Project. Courtesy of John T. Foley, Professor, Department of Physics and Astronomy, Mississippi State University, USA.
Cover production: *design & production* GmbH, Heidelberg

Printed on acid-free paper   SPIN: 11320623   57/3141/YU   5 4 3 2 1 0

Preface to *Progress in Nano-Electro-Optics*

Recent advances in electro-optical systems demand drastic increases in the degree of integration of photonic and electronic devices for large-capacity and ultrahigh-speed signal transmission and information processing. Device size has to be scaled down to nanometric dimensions to meet this requirement, which will become even more strict in the future. In the case of photonic devices, this requirement cannot be met only by decreasing the sizes of materials. It is indispensable to decrease the size of the electromagnetic field used as a carrier for signal transmission. Such a decrease in the size of the electromagnetic field beyond the diffraction limit of the propagating field can be realized in optical near fields.

Near-field optics has progressed rapidly in elucidating the science and technology of such fields. Exploiting an essential feature of optical near fields, i.e., the resonant interaction between electromagnetic fields and matter in nanometric regions, important applications and new directions such as studies in spatially resolved spectroscopy, nanofabrication, nanophotonic devices, ultrahigh-density optical memory, and atom manipulation have been realized and significant progress has been reported. Since nanotechnology for fabricating nanometric materials has progressed simultaneously, combining the products of these studies can open new fields to meet the above-described requirements of future technologies.

This unique monograph series entitled "Progress in Nano-Electro-Optics" is being introduced to review the results of advanced studies in the field of electro-optics at nanometric scales and covers the most recent topics of theoretical and experimental interest on relevant fields of study (e.g., classical and quantum optics, organic and inorganic material science and technology, surface science, spectroscopy, atom manipulation, photonics, and electronics). Each chapter is written by leading scientists in the relevant field. Thus, high-quality scientific and technical information is provided to scientists, engineers, and students who are and will be engaged in nano-electro-optics and nanophotonics research.

I gratefully thank the members of the editorial advisory board for valuable suggestions and comments on organizing this monograph series. I wish to express my special thanks to Dr T. Asakura, Editor of the Springer Series in Optical Sciences, Professor Emeritus, Hokkaido University for recommending me to publish this monograph series. Finally, I extend an acknowledgement to

Dr Claus Ascheron of Springer-Verlag, for his guidance and suggestions, and to Dr H. Ito, an associate editor, for his assistance throughout the preparation of this monograph series.

Yokohama, October 2002   *Motoichi Ohtsu*

Preface to Volume IV

This volume contains four review articles focusing on different aspects of nano-electro-optics. The first chapter reviews a versatile scanning near-field optical microscope with magnetic contrast by utilizing a Sagnac interferometer for monitoring the magneto-optical Kerr effect. This microscope is used to characterize data-storage media as well as to study the formation of microdomain patterns in ultrathin magnetic films.

The second chapter aims at describing how to achieve high-quality T-shaped quantum wires with high spatial uniformity. To characterize local structural and optical properties in quantum wires, a high-resolution microscopic photoluminescence method is used. Lasing from a single-quantum-wire laser structure is also demonstrated.

The third chapter summarizes material parameters of InGaN, and then general transition modes are discussed based on screening of the piezoelectric field, as well as on localization behavior of exciton/carriers. Detailed results are also shown on near-field luminescence mapping in InGaN/GaN single-quantum-well structures in order to interpret the recombination mechanism in InGaN-based nanostructures.

The last chapter concerns the theoretical treatments of optical near field and optical near-field interactions. The half-space problems are solved based on the angular-spectrum representation of the scattered field, where the energy transfer between interacting objects is made clear. This treatment provides the basis to investigate the signal transport and associated dissipation in nano-optical devices.

As was the case of Volumes I–III, this volume is published by the support of an associate editor and members of the editorial advisory board. They are:

Associate editor: Ito, H. (Tokyo Inst. Tech., Japan)
Editorial advisory board: Barbara, P.F. (Univ. of Texas, USA)
                                Bernt, R. (Univ. of Kiel, Germany)
                                Courjon, D. (Univ. de Franche-Comte, France)
                                Hori, H. (Univ. of Yamanashi, Japan)
                                Kawata, S. (Osaka Univ., Japan)
                                Pohl, D. (Univ. of Basel, Switzerland)
                                Tsukada, M. (Univ. of Tokyo, Japan)
                                Zhu, X. (Peking Univ., China)

I hope that this volume will be a valuable resource for the readers and future specialists.

Tokyo, July 2004   *Motoichi Ohtsu*

Contents

Quantum Theory of Radiation in Optical Near Field
Based on Quantization of Evanescent Electromagnetic Waves
Using Detector Mode

Tetsuya Inoue, Hirokazu Hori................................... 127

List of Contributors

Hidefumi Akiyama
Institute for Solid State Physics
University of Tokyo
5-1-5 Kashiwanoha, Kashiwa
Chiba 277-8581, Japan
golgo@issp.u-tokyo.ac.jp

Andreas Bauer
Institut für Experimentalphysik
Freie Universität Berlin
Arnimallee 14
14195 Berlin, Germany
bauer@physik.fu-berlin.de

Hirokazu Hori
Interdisciplinary Graduate School
of Medicine and Engineering
University of Yamanashi
4-3-11 Takeda
Kofu 400-8511, Japan
hirohori@yamanashi.ac.jp

Tetsuya Inoue
Department of Electronics
Yamanashi Industrial Technology
College
1308 Kamiozo
Enzan 404-0042, Japan
t_inoue
@swallow.elec.yitjc.ac.jp

Günter Kaindl
Institut für Experimentalphysik
Freie Universität Berlin
Arnimallee 14
14195 Berlin, Germany
kaindl@physik.fu-berlin.de

Akio Kaneta
Department of Electronic Science
and Engineering
Kyoto University
Katsura Campus, Nishikyo-ku
Kyoto 615-8510, Japan
kaneta@
fujita.kuee.kyoto-u.ac.jp

Yoichi Kawakami
Department of Electronic Science
and Engineering
Kyoto University
Katsura Campus, Nishikyo-ku
Kyoto 615-8510, Japan
kawakami@kuee.kyoto-u.ac.jp

Gereon Meyer
Institut für Experimentalphysik
Freie Universität Berlin
Arnimallee 14
14195 Berlin, Germany
meyerg@physik.fu-berlin.de

Takashi Mukai
Nitride Semiconductor Research
Laboratory
Nichia Corporation
491 Oka, Kaminaka, Anan
Tokushima 774-8601, Japan
tmukai@nichia.co.jp

Yukio Narukawa
Nitride Semiconductor Research
Laboratory
Nichia Corporation
491 Oka, Kaminaka, Anan
Tokushima 774-8601, Japan
narukawa@hq.nichia.co.jp

Kunimichi Omae
Paul-Drude-Institut
für Festkörperelektronik
Hausvogteiplatz 5-7
10117 Berlin, Germany
komae@pdi-berlin.de

Masahiro Yoshita
Institute for Solid State Physics
University of Tokyo
5-1-5 Kashiwanoha, Kashiwa
Chiba 277-8581, Japan
yoshita@issp.u-tokyo.ac.jp

Near-Field Imaging
of Magnetic Domains

Gereon Meyer, Andreas Bauer, and Günter Kaindl

1 Introduction

The imaging of magnetic domains is of high technological interest for the characterization of data-storage media and nonvolatile memory devices, where domains carry the bit information [1]. It also has an impact on the study of magnetism in general, since the formation of domain patterns in thin magnetic films is currently a field of intense research. For the observation of ultrafast switching processes in submicrometer-size domains a magnetic microscope is needed with picosecond temporal and nanometer lateral resolution that allows the application of external magnetic fields.

Quite a few powerful methods for the imaging of magnetic domains have been introduced in the past, but most of them do not fulfill the described prerequisites: Spin-polarized scanning tunneling microscopy (SP-STM) can distinguish the spin orientation of adjacent atoms [2]; its temporal resolution, however, remains in the range of milliseconds. Other methods with high lateral resolution face this drawback as well, like, e.g., magnetic force microscopy (MFM) [3] and scanning electron microscopy with polarization analysis (SEMPA) [4]. Magneto-optical microscopy in contrast can follow magnetization processes running on the picosecond time scale and below [5], but the lateral resolution is diffraction limited to about half the wavelength of light. The best compromise between spatial and temporal resolution is presently provided by magnetic X-ray microscopy [6] and photoemission electron microscopy (PEEM) [7] allowing picosecond temporal resolution together with the imaging of structures as small as a few tens of nanometers. The study of magnetization-reversal processes, however, requires the application of external magnetic fields, a condition that, in general, can only be met by optical methods and X-ray techniques (except for some very tricky setups with electron-microscopic techniques). A major challenge is thus to increase the lateral resolution of magneto-optical microscopes.

In this chapter we shall review our effort to build a versatile scanning near-field optical microscope (SNOM) with magnetic contrast by utilizing a Sagnac interferometer for monitoring the magneto-optical Kerr effect [8,9]. This new microscope (UHV-Sagnac-SNOM) has been used to characterize data-storage media [10] as well as to study the formation of microdomain patterns in ultrathin magnetic films [11,12]. For in situ studies, the SNOM can be operated in ultrahigh vacuum (UHV) [13].

2 Magneto-Optical SNOM

The measurement of magneto-optical effects with a SNOM poses some serious difficulties, since the changes in light polarization to be measured can be strongly affected by phenomena other than magnetism. Such artifacts, caused, e.g., by birefringence or topography, can be avoided by a careful choice of the near-field probe and, in particular, by using a Sagnac interferometer for detecting the magneto-optical Kerr effect. Therefore, we shall first introduce some basics of magneto-optics before reviewing previous magneto-optical SNOM work.

2.1 Faraday Effect and Kerr Effect

When linearly polarized light interacts with a magnetic sample, it is converted into elliptically polarized light, with the main axis of the ellipse being rotated by a certain angle [14]. This magneto-optical effect is called the Faraday effect in transmission and the Kerr effect in reflection. The rotation angle and the ellipticity are expressed by the real and the imaginary parts of a complex angle ϕ_F (resp. ϕ_K) (see Fig. 1). To a first approximation, both parts are proportional to the sample magnetization, M, and can hence be used to monitor the response of magnetization to variations of parameters like temperature or external magnetic field. The recording of hysteresis loops by measuring the magneto-optical Kerr effect (MOKE) versus the external magnetic field is now a standard procedure to study magnetization-reversal processes. In combination with pulsed laser sources, magnetization dynamics can be studied by magneto-optics down to the subpicosecond time scale. It also provides a powerful method for imaging magnetic domains. Technical applications include Faraday isolators in optical communication and the Kerr-effect-based readout process in magnetic data storage.

Magneto-optical effects are caused by the interplay of magnetic circular dichroism and magnetic circular birefringence: Linearly polarized light is a superposition of left-handed and right-handed circularly polarized light (LCP and RCP). Dichroism means a difference in the absorption coefficients of the two components, while birefringence results from a difference in the phases. The material looks different for LCP and RCP, because M is an axial vector. Any symmetry transformation that converts LCP into RCP changes the orientation of M. Dichroism and birefringence are related to each other by the Kramers–Kronig relations. In general, the combination of LCP and RCP – after interaction – results in ellipticity *and* rotation of the light polarization vector (Fig. 2). In magnetic transition metals the rotation angles are rather small, with typical values amounting to only a few tenths of a degree [15].

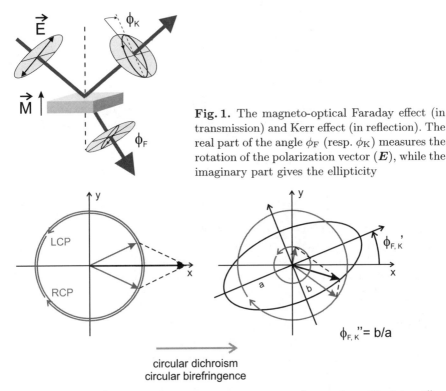

Fig. 1. The magneto-optical Faraday effect (in transmission) and Kerr effect (in reflection). The real part of the angle ϕ_F (resp. ϕ_K) measures the rotation of the polarization vector (E), while the imaginary part gives the ellipticity

circular dichroism
circular birefringence

Fig. 2. The rotation angle of the polarization vector, $\phi'_{F,K}$, and its ellipticity, $\phi''_{F,K}$, result from a combination of magnetic circular dichroism and circular birefringence

These effects are microscopically understood by a preference of one kind of spin over the other in optical excitations, which is caused by a combination of exchange splitting and spin-orbit coupling [14]. In the framework of a phenomenological theory, the magnetization induces complex nondiagonal elements of the dielectric tensor ϵ [16].

In both cases, domain imaging and recording of magnetization curves, a signal proportional to M is needed. Such a signal follows from the magneto-optical effects if a setup with crossed polarizers is used. For this purpose, a polarizer is inserted into the reflected beam (for Kerr-effect detection) with the main axis slightly misaligned from extinction by an angle α. The intensity of transmitted light, I, is then given by:

$$I(\pm M) = I_0 \sin^2(\alpha \pm \phi_K) + I_B , \qquad (1)$$

where $\pm M$ is the magnetization that causes a Kerr rotation by an angle ϕ_K[1]; I_0 is the intensity of the reflected light, and I_B is the background intensity

[1] Complex and real parts of ϕ_K need not be distinguished in this consideration, since both are proportional to M.

due to the limited extinction ratio of the polarizer, $\epsilon = I_B/I_0$. The magneto-optical contrast C is defined by

$$C = \frac{I(+M) - I(-M)}{I(+M)} = \frac{I_0(\sin^2(\alpha + \phi_K) - \sin^2(\alpha - \phi_K))}{I_0\sin^2(\alpha + \phi_K) + I_B} . \quad (2)$$

For small angles α and ϕ_K slightly misaligned from extinction ($\alpha \gg \phi_K$), this gives a signal proportional to ϕ_K, which is used as a measure of M:

$$C \approx \frac{4\alpha\phi_K}{\alpha^2 + \epsilon} . \quad (3)$$

2.2   Sagnac Interferometer

A Sagnac interferometer is a sensor for testing time-reversal symmetry [17]. It measures the phase shift, $\Delta\phi$, between two light beams that propagate in opposite directions through the interferometer. Such a phase shift can result from nonreciprocal propagation conditions due to broken time-reversal symmetry. Mechanical rotation of the optical path, e.g., breaks time-reversal symmetry due to a relativistic effect that causes different optical pathlengths for clockwise (CW) and counterclockwise (CCW) light beams. The Sagnac effect is thus used in inertial sensors like laser gyroscopes to detect the angular frequency of, e.g., an aircraft. Magneto-optical effects are another example. If the counterpropagating beams are converted into circularly polarized light, the Sagnac interferometer can measure their phase shift upon interaction with the magnetic sample, i.e., the magneto-optical rotation [18]. We shall show further that this may be the most appropriate way of monitoring the Kerr effect in SNOM.

The principle of a glass-fiber-based Sagnac interferometer is shown in Fig. 3: The beam of a laser source is divided into two partial beams by a 50:50 beam splitter, one traveling clockwise (CW), the other traveling counterclockwise (CCW) through a glass-fiber loop. Finally, both beams interfere at a photo detector. Any nonzero $\Delta\phi$ between the corresponding electric fields E_{CW} and E_{CCW} reduces the intensity of the interference signal, $I(\Delta\phi)$, given by

$$I = \frac{1}{2}\left|E_{CW}e^{i\Delta\phi} + E_{CCW}\right|^2 = \frac{1}{2}I_0(1 + \cos(\Delta\phi)) . \quad (4)$$

Here, we assume that $E_{CW}^2 = E_{CCW}^2 = 1/2I_0$ with I_0 being the output intensity of the laser source. And, we take into account the fact that half of the light intensity exits the interferometer via the nonreciprocal port.

To increase the sensitivity of the interferometer and to distinguish polarities of $\Delta\phi$, a phase modulator is used. Its refractive index oscillates at a high frequency, ω, modulating the phase of the light beam as $\phi(t) = \phi_m\sin(\omega t)$. This modulation is most effective if the two partial beams reach the phase

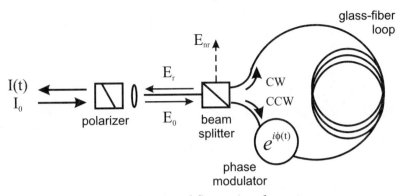

Fig. 3. Setup of a glass-fiber-based Sagnac interferometer

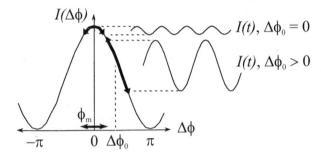

Fig. 4. Sagnac-interference signal $I(\Delta\phi)$ with phase modulation at amplitude ϕ_m and frequency ω. The ω signal indicates a nonreciprocal phase shift, $\Delta\phi_0$

modulator delayed by $\tau = \pi/\omega$, which means $\phi(t - \tau) = -\phi(t)$. The delay, $\tau = nL/c$, can be adjusted by proper choice of fiber length, L, and refractive index, n. The detected intensity then amounts to:

$$I = \frac{1}{2}\left|E_{CW}e^{i\phi(t)+i\Delta\phi} + E_{CCW}e^{i\phi(t-\tau)}\right|^2 \tag{5a}$$

$$= \frac{1}{2}I_0(1 + \cos(2\phi_m\sin(\omega t) + \Delta\phi)) \, . \tag{5b}$$

A first-harmonic (ω) oscillation of I indicates that $\Delta\phi \neq 0$, whereas the second-harmonic (2ω) part exists also for $\Delta\phi = 0$, representing a good test for the proper alignment of the interferometer (Fig. 4). This can also be concluded from an expansion of I with respect to its frequency components, where J_n are Bessel functions:

$$\frac{I}{I_0} = \frac{1}{2}\left[1 + \cos(\Delta\phi)J_0(2\phi_m)\right] \tag{6a}$$

$$- \left[\sin(\Delta\phi)J_1(2\phi_m)\right]\sin\omega t \tag{6b}$$

$$+ \left[\cos(\Delta\phi)J_2(2\phi_m)\right]\cos 2\omega t \tag{6c}$$

$$+ \ldots$$

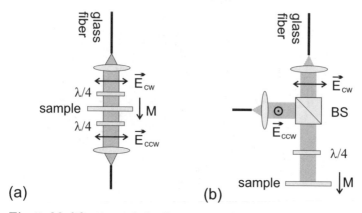

Fig. 5. Modifications of the Sagnac interferometer to measure (**a**) the magneto-optical Faraday effect and (**b**) the Kerr effect

The modifications of the Sagnac interferometer that are necessary to measure the magneto-optical effects are shown in Fig. 5: The counterpropagating beams are converted into circularly polarized light with the same helicity for Faraday-effect measurements (Fig. 5a) and with opposite helicity for Kerr-effect measurements (Fig. 5b). In both cases, the helicity is parallel to M for one beam and antiparallel for the other one. The magneto-optical effects induce a phase shift, $\Delta\phi$, since the refractive indices of the CW and CCW beams are different due to the magnetic circular birefringence. $\Delta\phi$ equals twice the rotation angle of the polarization vector, and it changes sign upon magnetization reversal, which can be measured by phase-sensitive detection of the ω signal.

The main difference between a Sagnac interferometer and the crossed-polarizers method is the exclusive sensitivity of the Sagnac signal to nonreciprocal effects. Polarization changes caused by effects other than magneto-optical ones do not contribute to the Sagnac signal but would be measured by the crossed-polarizers method as well. This is important for magnetic-domain imaging, where additional polarization changes due to, e.g., optical activity of the sample (or, for SNOM, of the tip) can be much stronger than the magneto-optical contrast. Optical activity, however, does not break the time-reversal symmetry. This can readily be seen, e.g., by comparing transmission of linearly polarized light by a perpendicularly magnetized sample and a $\lambda/2$-retardation plate: The latter rotates the polarization vector of light propagating along its main axis always in the same sense, whatever the propagation direction is. A magnetic material, in contrast, rotates the polarization vector clockwise for parallel orientation of propagation and magnetization vectors, and counterclockwise for the antiparallel orientation.

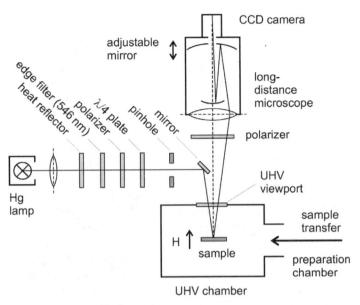

Fig. 6. UHV-Kerr microscope optimized for magneto-optical contrast rather than for lateral resolution

2.3   Kerr Microscopy

A far-field Kerr microscope is a standard tool for the imaging of magnetic domains, since it provides strong magnetic contrast using a conventional light source [1]. In contrast to electron microscopes, this contrast is not affected by external magnetic fields – a necessity if magnetization-reversal processes are studied.

The setup of a Kerr microscope is rather simple: Linearly polarized light is focused onto the sample and the illuminated area is imaged by a conventional microscope equipped with a second (crossed) polarizer in front of the objective lens. To avoid diffraction and interference, an incoherent light source is used, like a Hg lamp, providing sufficient intensity for measurements with almost crossed polarizers.

The performance of a Kerr microscope is determined by two parameters: the resolving power and the magneto-optical sensitivity, which limit each other. According to the Rayleigh criterion, lateral resolution can be increased using a higher numerical aperture. But raising the numerical aperture reduces the magneto-optical contrast due to the widening of the acceptance angle. This dilemma is particularly important for the application of Kerr microscopy to in situ studies of magnetic domains in ultrathin films. Such a microscope is usually placed in front of a viewport of the UHV chamber at a long distance from the sample, which limits the numerical aperture (see the UHV setup schematically shown in Fig. 6). In addition, the domain contrast of ultrathin films is weak. Most UHV-Kerr microscopes have therefore been optimized for contrast rather than for resolution, which is a few micrometers [19]. However, high-resolution setups have been built as well [20]. An alternative method is a scanning Kerr microscope, where the sample is probed by a focused laser beam in a confocal setup [21]. An interesting application of this method is the time-resolved observation of ultrafast magnetization dynamics in a pump-probe experiment [5].

2.4   Domain Contrast in SNOM

Neither the bits of state-of-the-art data-storage media nor the stripe-domain patterns expected in ultrathin magnetic films can properly be imaged by far-field Kerr microscopy. At first glance, methods like MFM or SEMPA seem to be more appropriate to study such submicrometer-scale magnetic domains. But magneto-optics has two major advantages: Compatibility with external magnetic fields and the potential to combine it with pulsed laser sources for the study of ultrafast magnetization dynamics. The diffraction-limited lateral resolution, being the main drawback of magneto-optics, should be avoidable in a scanning near-field optical microscope (SNOM) [22–24].

A simple setup of a magneto-optical SNOM is shown in Fig. 7: A probe for simultaneous emission and collection of light is scanned in close proximity ($\ll \lambda$) across the sample surface. Typically, it is a glass-fiber tip covered by a metal film. A tiny hole in the metal (approximate diameter: 100 nm) acts as an aperture. On illumination, the near field of the tip interacts with the sample material within an area that is limited by the aperture size. At each scanning position, all reflected light originates from this laterally limited near-field excitation. By insertion of crossed polarizers, the magneto-optical Kerr effect can locally be measured giving a highly resolved magnetic-domain image. The corresponding topography can be visualized by plotting the deflection of the tip that is necessary to maintain a constant tip-to-sample distance. Proper interpretation of the optical image requires comparison with topography, because crosstalk can induce additional contrast in the optical image – a typical SNOM artifact [25,26].

The first demonstration of magneto-optical contrast by SNOM was published by Betzig et al. [27]. They coupled the linearly polarized light of an Ar^+ laser into a glass fiber that was tapered and covered by a metal film to form a small aperture. Using a crossed-polarizers setup, they determined the Faraday rotation of light that was transmitted by the sample upon illumination by the near field of the tip. They succeeded in imaging magnetic domains of perpendicularly magnetized Co/Pt-multilayer films, and they claimed a rather spectacular lateral resolution of 30–50 nm. This work entailed quite a lot of magneto-optical SNOM experiments. Most of them were done in transmission [28–35].

Only very few instruments have been built to image magnetic domains in reflection by measuring the magneto-optical Kerr effect (Fig. 8). Two types

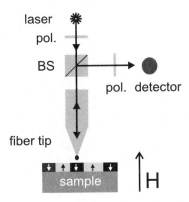

laser

pol.

BS

pol. detector

fiber tip

sample

H

Fig. 7. Combination of shared-aperture SNOM and magneto-optical Kerr effect for magnetic-domain imaging (BS: beam splitter, H: external magnetic field, pol.: polarizer)

of near-field probe have successfully been used for this purpose so far: The aperture probes [36,37], and apertureless probes being excited to light emission, i.e., either a metallic nanoparticle [38] or a cantilever-shaped tungsten tip [39].

The aperture-type setups (see, e.g., Fig. 8a) used metal-coated glass-fiber tips for illumination of the sample with linearly polarized light (of different wavelengths $\lambda = 635$ nm [36], $\lambda = 488/512$ nm [37]). The reflected light was collected by a spherical mirror. It passed a polarizer and was focused onto a photodetector located behind the sample. A photoelastic modulator (PEM) was applied to increase the magneto-optical sensitivity. In one setup, the Kerr rotation angle could be determined quantitatively [37]. Magnetic domains that had been written thermomagnetically into a Co/Pt-multilayer film with perpendicular easy axis and $0.25°$ Kerr rotation[2] could be imaged. The signal-to-noise ratio, however, was rather weak (1:1 [36] and 4:1 [37]). The lateral resolution[3] amounted to 50 nm [36] and 200 nm [37].

The first SNOM measuring the Kerr effect used a metal particle as a scattering probe [38]. A silver globule with a mean diameter of 40 nm was glued onto the surface of a hemispherical glass body. It is illuminated by the evanescent field of a totally reflected laser beam within the glass body. When optically excited at the surface-plasmon frequency, such a particle radiates like an oscillating dipole (Fig. 8b). The magneto-optical Kerr effect was measured by insertion of a polarizer into the perpendicularly scattered light. By this technique, it was possible to image magnetic domains with a diameter of 0.5 µm in a perpendicularly magnetized Co/Pt-multilayer film ($0.15°$ Kerr rotation) at a signal-to-noise ratio of 5:1. The lateral resolution was 200 nm. It should be noted that it was even possible to image magnetic domains in a permalloy film, where the magnetization vector was lying in the film plane. One would usually not expect any contrast of that sample in

---

[2] We always understand Kerr rotation as half of the difference between the rotation angles corresponding to oppositely magnetized domains.

[3] Lateral resolution is understood as the total width of domain walls.

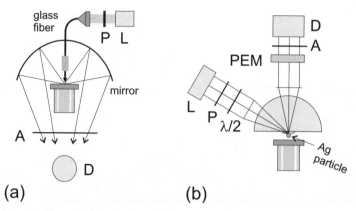

Fig. 8. Kerr-SNOM setups: (a) aperture type and (b) apertureless type. L: laser, P: polarizer, A: analyzer, BS: beam splitter, D: detector

the polar Kerr-effect geometry used here. The authors assume that it was a pure near-field phenomenon, since the contrast exponentially decayed upon increasing the tip-to-sample distance [40].

Apertureless SNOM is usually based on the scattering of the electromagnetic field by the tip of a metallic cantilever as it is used in an atomic force microscope (AFM). Such a setup has been combined with magneto-optical Kerr-effect detection as well [39]: The sample was illuminated by the focused beam of a laser diode. Light scattered by the tip was collected by a microscope objective. Near-field and far-field components of the reflected light can be distinguished using a lock-in amplifier, since the near-field contributions are modulated by the oscillation frequency of the tip. Using this technique, it was possible to image magnetic domains of a Co/Pt-trilayer film with 0.1° Kerr rotation. The authors claim a signal-to-noise ratio of 4:1 and a lateral resolution of 200 nm.

The most convincing performance was obtained by the scattering-particle-type Kerr SNOM [38]. The strongest limitation comes from nonmagnetic polarization changes like birefringence due to mechanical strain of the probe or the sample (i.e., optical activity). With aperture-type SNOM, an additional problem is that scanning of a metal-surrounded aperture across a conducting surface affects the polarization state of the probe [41]. The polarization contrast can also be influenced by the topographic structure of the sample [42].

At first glance, apertureless Kerr SNOM seems to be the better choice. It has also been studied theoretically in more depth [43,44]. But low intensities and strong nonmagnetic backgrounds lead to limitations as well. We therefore decided to follow the path of aperture-type Kerr SNOM, but tried to avoid some of the typical sources of trouble: We used uncoated glass-fiber tips instead of metal-coated ones. Resolution of better than $\lambda/3$ has already been demonstrated [45], and the optical throughput is much higher for uncoated fiber tips. In addition, we took up an idea published by Kapitulnik et al. [46] to replace the crossed polarizers by a Sagnac interferometer.

3   Experimental Details

The intention of our work was to apply SNOM to in situ studies of magnetic domains in ultrathin films at various temperatures, film thicknesses and magnetic fields. The high spatial resolution of the near-field method should allow for imaging microstructures that cannot be resolved by far-field Kerr microscopy. The latter provides overview images on a larger length scale. For thin-film studies, we first used SNOM in UHV. Then, we combined it with a Sagnac interferometer for Kerr-effect detection. The Sagnac signal also provides an easy way of optical tip-to-sample-distance control, which in UHV is superior to the common shear-force method.

3.1   UHV System

UHV is the standard experimental environment for surface physics. It slows down the rate of adsorption of residual-gas molecules onto substrate surfaces, which is a prerequisite for epitaxial growth of ultrathin metal films. The sample could also be covered by a protective coating for studies in air, but such coatings might affect the magnetic anisotropies of the film. This is undesirable for investigations of processes depending strongly on the anisotropies, like domain formation.

In our setup, the UHV chamber is connected to two further chambers: a preparation chamber for ultrathin films equipped with standard techniques for preparation of single-crystal surfaces, film growth, and characterization, and a chamber containing a room-temperature scanning tunneling microscope (STM) for studies of the film morphology. The base pressure is 5×10^{-11} hPa, and during film preparation the pressure does not rise beyond 1×10^{-10} hPa.

After preparation, the ultrathin-film samples are transferred for in situ experiments into the chamber for magneto-optical studies (Fig. 9), where they are put into a sample cage made of Cu, which is mounted directly onto the cold finger of the cryostat. For proper thermal contact, the sample is pressed by a spring onto the walls of the cage. A 5-mm bore in the cage wall makes the film surface accessible for optical studies. The cryostat is equipped with a Si diode for precise determination of the sample temperature. By regulation of the LHe flow and by heating, the sample temperature can be adjusted in the range between about 20 K and 450 K.

The chamber is equipped with three magneto-optical devices differing in lateral resolution and Kerr-effect sensitivity: The first one allows the measurement of polar and longitudinal MOKE that indicates the net magnetization behavior of the sample laterally averaged over the area illuminated by the laser spot (≈ 1 mm^2). The second device is a far-field UHV-Kerr microscope

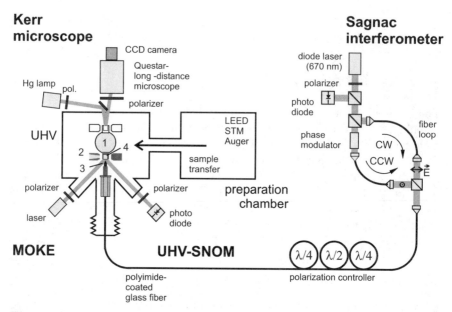

Fig. 9. UHV system for magnetic microscopy of ultrathin films equipped with MOKE, UHV-Kerr microscope and UHV-Sagnac-SNOM. (1) LHe-flow cryostat, (2) UHV electromagnet, (3) transferable SNOM head, and (4) sample cage

that allows an overview image of the domain structure of the sample to be taken. Being optimized for magneto-optical sensitivity, our setup (Fig. 6) provides a lateral resolution of 3 µm. The third device is a magneto-optical SNOM measuring the Kerr effect via a Sagnac interferometer (UHV-Sagnac-SNOM). At present, the lateral resolution is one order of magnitude better than that of the Kerr microscope, so that one can study the domain structure in more detail. Switching between these methods requires no further sample transfer, just a 180° rotation of the cryostat to bring the sample into the focus of the Kerr microscope. In this way, we can successively apply all three methods at fixed sample temperature. For proper positioning, the cryostat is mounted onto a rotary feedthrough in combination with a jacking stage and an x-y stage (Fig. 10). All optical components are located outside the chamber, just a glass fiber is fed through a bore in an UHV flange to connect the UHV-SNOM head with the Sagnac interferometer, which is placed on an optical table next to the UHV chamber.

The electromagnet that is used to induce magnetization reversal is located inside the UHV chamber. It is rotatable, and the sample cage can be placed between the pole shoes of its iron yoke. In this way, external magnetic fields ranging up to 1500 Oe can be applied either in the film plane or perpendicular to it. One of the pole shoes has a conical bore that guides the light to the sample for polar-MOKE measurements.

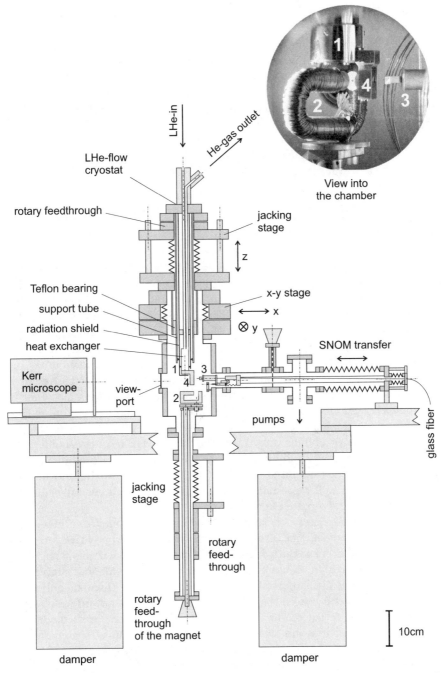

Fig. 10. Drawing of the UHV chamber for magneto-optical microscopy, equipped with a LHe-flow cryostat and an electromagnet. For (1–4) see the caption of Fig. 9

Vibrations of tip and sample can influence the optical signal, because the near-field intensity depends critically on the distance. This can be suppressed, like in STM, by a combination of stages with strongly different resonance frequencies. We mounted the entire UHV system onto pneumatic dampers with a resonance frequency of 2 Hz. A rigid tip–sample configuration would be the best second damping step, but tip and sample should be separately transferable in our UHV-Sagnac-SNOM. We therefore set aside part of the desired rigidity and tried to compensate for this by making all oscillating parts as short as possible: Upon transfer, the SNOM head is put into a holder that is directly fixed to the inner chamber wall. The cryostat is additionally stabilized by a rigid stainless steel support tube with a Teflon bearing at its end.

3.2   UHV-SNOM Setup

The transferable SNOM head consists of two parts: a piezo tube for $x–y$ scanning and z-approach and a dither unit, necessary for the control of the tip-to-sample distance (Fig. 11). The SNOM tip is used here both for illumination and for collection of the reflected light (shared-aperture mode), which is the easiest way of realizing a reflection-mode SNOM in combination with a Sagnac interferometer. In this way, no further optical components except for the glass fiber are needed inside the UHV chamber. The 670-nm light beams (CW and CCW) are guided by a single-mode fiber (with a cutoff wavelength of 580 nm) coated by a polyimide layer. This coating has a low outgassing rate and is therefore compatible with the bakeout procedure that is required to obtain ultrahigh vacuum. The tip is formed by etching the glass fiber in hydrofluoric acid (HF) according to the Turner method [47]. The more convenient tube-etching method [48] cannot be applied to these fibers since it induces asymmetric etching due to the fact that the HF-inert polyimide coating does not sit equally tight on the glass fiber. Removal of the polyimide is done by dipping the fiber into boiling NaOH (30%) for approximately 60 min. Good tips have a well-reproducible cone angle of 23°, and the far-field light spot shows a characteristic circular shape.

All experiments described here were performed with uncoated glass-fiber tips, which provide a higher throughput and better defined polarization properties than coated ones. Both kinds of tips differ in the way light is guided to the near-field region. Glass fibers lose the ability to guide and confine the fundamental mode if the diameter of the core gets smaller than one half of the wavelength. In metal-coated tips, the light has to couple to a surface plasmon of the metal film to get to the aperture [49]. The efficiency of such coupling is usually low, which is the reason for the low throughput of coated tips. In uncoated tips, the light can reach the extreme tip by internal total reflection. Lateral resolution is lower, however, because part of the light intensity leaks out before arriving at the extreme tip. Uncoated tips are therefore modeled by a combination of a subwavelength aperture and a far-field aperture [50]. There is a certain controversy over the question of whether it is actually

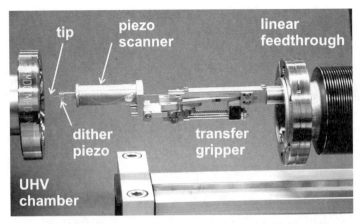

Fig. 11. UHV-SNOM head with gripper for transfer into a garage, where tips can be changed

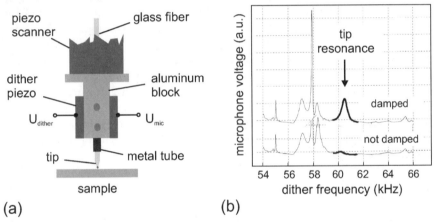

Fig. 12. (a) SNOM head setup for tip–sample distance control and scanning in UHV. (b) Resonance spectra with and without damping of tip oscillation

possible to beat the diffraction limit using uncoated glass-fiber tips. Samples with strong topographic features could be imaged with apparently high lateral resolution using uncoated tips. However, this did not come from any near-field sensitivity, but simply from crosstalk. Such artifacts can largely be excluded for our samples, which are topographically flat single-crystal surfaces. True lateral resolution almost at the diffraction limit (in the range of $\lambda/2$ to $\lambda/3$) has already been observed using uncoated tips [25,45,51].

We tested two different approaches for tip–sample distance control in UHV: the conventional shear-force method [52] and an optical method. A dither unit to apply the shear-force method in UHV was designed. It consists of an Al block that holds the glass fiber supported by a metal tube and has piezo plates on opposite sides (see Fig. 12a). One of these plates is used to excite horizontal tip oscillations, while the second plate acts as a microphone to monitor the oscillations of the block. These are particularly strong at resonance frequencies of any single component of the setup. The amplitude of tip oscillations (and their resonance peak) gets strongly damped upon approach to the sample, which is why it can be used as the controlled quantity in a feedback circuit.

The resonance frequency of the tip is just the eigenfrequency ν of a bending rod, given by:

$$\nu = 0.28 \frac{R}{L^2} \sqrt{\frac{E}{\rho}} \, , \tag{7}$$

with L being the length of the tip, R the radius, E the Young's modulus, and ρ the density. This amounts to about 60 kHz in the present case. In contrast to the tuning-fork concept, ν depends strongly on the length of the fiber tip and will hence differ slightly from tip to tip even with careful control of the lengths of the tips. The exact value can be found by a comparison of spectra with and without damping (Fig. 12b). Bakeout procedures and variation of sample temperature do not have any noticeable influence on the resonance frequencies. Such effects, however, have been reported for other setups, where both sample and SNOM head were cooled to low temperature [53]. Even though all our studies of ultrathin films were done using the optical distance control (described in the next paragraph), the shear-force distance control was additionally run for safety reasons: to protect the tip if the optical distance control should fail.

3.3   Sagnac-SNOM Setup

Another unique feature of the present SNOM setup, besides UHV operation, is the detection of the Kerr effect via a Sagnac interferometer. The type of Sagnac interferometer used here has been described by Spielman [18], who had applied it to magneto-optical studies as well, but only in the far field. As light source, a laser diode (670 nm) is used. It has a relatively short coherence length, necessary to avoid competing interferences induced by reflections from the surfaces of optical components. The output power of the laser, after passing a Faraday isolator and beam-shaping optics, is 1.8 mW.

The setup of the Sagnac-SNOM is shown in more detail in Fig. 13. A key conceptual decision is the proper choice of the phase-modulation frequency ν_{mod} at the electro-optical modulator. It has to correspond to the relative delay τ between the counterpropagating beams. According to Sect. 2.2, optimal tuning requires

$$\nu_{mod} = \frac{1}{2\tau} = \frac{c}{2nL} \, . \tag{8}$$

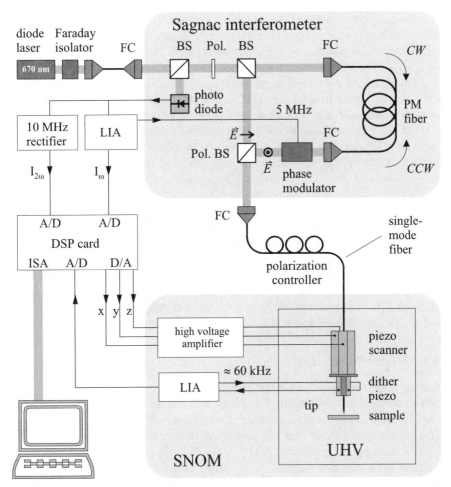

Fig. 13. Optics of the UHV-Sagnac-SNOM and electronics for distance control and data recording. FC: fiber coupler, Pol: polarizer, BS: beam splitter. The polarization controller consists of three fiber loops that act as an adjustable combination of two $\lambda/4$ plates and one $\lambda/2$ plate

For a fiber-loop length of $L = 20$ m and a refractive index of $n = 1.48$, this gives a modulation frequency of $\nu_{\mathrm{mod}} \approx 5$ MHz. The ac signal of a function generator is tuned to this frequency and used to excite a resonant electro-optical modulator (New Focus PM 4001); the corresponding TTL-signal serves as reference for lock-in detection.

We use a photodiode as detector, which is contained inside an aluminum box and is battery operated to minimize noise. Its output signal is analyzed with respect to the amplitudes of both the ω and 2ω signal (here $\omega = 2\pi\nu_{\mathrm{mod}}$). The 2ω signal represents the amount of light that has interacted with the sample and contributed to the interference signal. To determine its amplitude for distance control, a 10-MHz rectifier is applied. The ω signal, in contrast, carries the pure magneto-optical information, it is measured simultaneously by a lock-in amplifier.

The feedback routine is implemented as a program written in C and runs on a digital-signal-processing (DSP) card. It reads in three signals: (a) the output voltage of the lock-in amplifier measuring the amplitude of the ω part of the Sagnac-interference signal; (b) the output voltage of the 10-MHz bandpass filter representing the 2ω signal; (c) the output voltage of a second lock-in amplifier that monitors the amplitude of the tip oscillations, which is a measure of the shear forces. While signal (a) forms the magneto-optical image, signals (b) and (c) are used for distance control. They are compared to the set values of total intensity and tip damping, respectively, which correspond to certain tip-to-sample distances. The deviation from those set values determines the output voltages that are applied to the piezo scanner to adjust the tip-to-sample distance. The optical distance control turned out to be more appropriate for thin-film studies in UHV. Therefore, we chose the set values of signals (b) and (c) in such a way that the shear-force control just steps in if the optical control fails, e.g., at a strongly light-absorbing surface defect. It then simply retracts the tip to avoid a crash.

3.4   Performance Tests

During construction of the UHV-Sagnac-SNOM we faced several open questions: Would the stability of the setup be sufficient to allow for optical or shear-force distance control without strong crosstalk? Would it really be possible to achieve submicrometer lateral resolution using uncoated glass-fiber tips? And, would the sensitivity of the UHV-Sagnac-SNOM actually be sufficient to allow domain imaging of ultrathin films? We therefore carried out a series of performance tests, before we tried to image unknown domain structures.

We first tested the shear-force distance control in UHV. We found that it works, but it behaves differently from in air: damping in UHV arises mainly from direct contact of tip and sample, as has been reported before for SNOM operation at low temperatures [54] or in vacuum (not UHV) [55]. We did not observe the action of a damping medium between tip and sample in vacuum, reported by other groups [56]. Hence, the danger of tip crashes is quite strong in our less-compact setup, where vibrations can amount to about ± 10 nm. This, however, does not affect optical distance control, where longer tip-to-sample distances can be chosen.

To test the shear-force distance control in UHV, we scanned a calibration sample: a grid of parallel lines with a periodicity of 3 μm and a depth of 100 nm etched into a SiO_2 layer on top of a Si wafer. The grooves were a little wider than the lands, as can be clearly seen in the topographical image

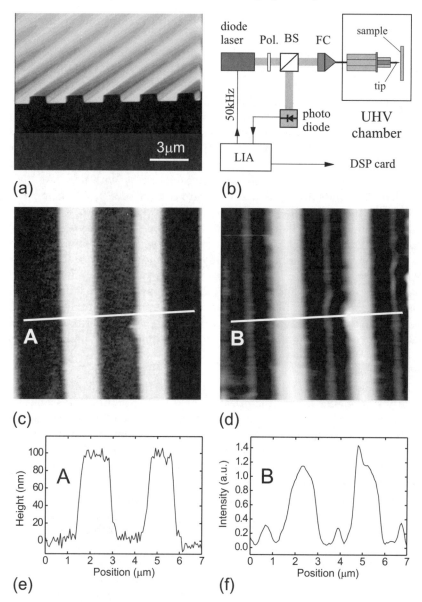

Fig. 14. (**a**) Calibration grating with 100-nm deep grooves etched into a SiO_2 layer on top of a Si wafer (image obtained with a scanning electron microscope, NT-MDT Inc., Moscow); (**b**) simple UHV-SNOM setup; (**c**) topographical image and (**d**) optical image obtained by UHV-SNOM; (**e, f**) contour plots along the lines A and B

shown in Fig. 14c. From the contour plot shown in Fig. 14e we derived a rms-noise amplitude of < 5 nm.

The same calibration sample was then also used to characterize the optical contrast that can be obtained with uncoated glass-fiber tips. We built a simple UHV-SNOM for that purpose (Fig. 14b): The light of the 670-nm diode laser is coupled into a tapered glass fiber, which is used as a SNOM tip in shared-aperture mode. The total reflected intensity is focused onto a photodiode. Figure 14d shows the optical image corresponding to the topographical image shown in Fig. 14c. The strong alternating contrast fits to the locations of lands and grooves with presumably very different reflectivities. There are some additional features that are typical for microscopy with uncoated glass-fiber tips: The optical contrast corresponding to defects in the topography is shifted in the optical image with respect to its position in the shear-force image. We think that such displacements arise from slightly different positions of the shear-force sensing tip (topography sensor) and the center of the optical sub-λ aperture, which might result from a certain asymmetry of the core axis or the etching procedure. This situation is analyzed in detail in Fig. 15: The topography sensor and the light-collecting aperture are shifted in such a way that the aperture passes the defect a few scan lines earlier than the topography sensor. The narrow light stripes running parallel to the grating lines in the center of the grooves are another artifact. They probably result from an interference effect.

To test the magneto-optical sensitivity of the UHV-Sagnac-SNOM, we imaged the magnetic bits of a magneto-optical disc (MOD). This sample is particularly suitable for the characterization of scanning-probe microscopes due to its pronounced topography and its strong, well-known magnetic-domain structure. This can help to detect crosstalk between topographical and magnetic contrast. We obtained the sample studied from Philips Research, Eindhoven. It was made of a TbFeCo alloy, i.e., a composition of transition and rare-earth metals, which is particularly suitable for data storage due to its perpendicular easy axis [57]. Its Kerr rotation is approximately 0.4° for red light. The surface of a MOD has a characteristic topography consisting of 1-μm wide lands with 600-nm wide and 100-nm deep grooves in between. The magnetic bits are approximately 1 μm wide and 3 μm long. They were written thermomagnetically into the tracks, i.e., the magnetization was reversed by local heating and application of a magnetic switching field. A first trial to image the magnetic bits by a Kerr SNOM using the crossed-polarizers setup failed because nonmagnetic polarization changes at the contour of the surface and at defects obscured the magnetic-domain pattern. This changed when the UHV-Sagnac-SNOM was used. Figure 16 shows topographical images (a,c) and the corresponding magneto-optical images (b,d). Lands and grooves can clearly be distinguished in the topography showing that shear-force distance control works well. The slight ripple structure arises from acoustically excited vibrations of tip and sample. From the contour line (Fig. 16e) we estimate the rms-noise level to be < 10 nm.

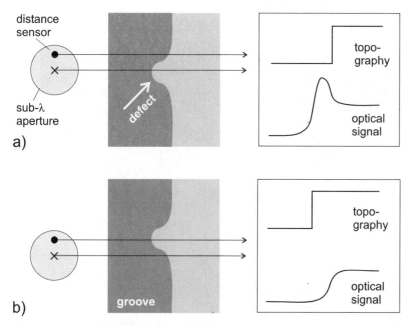

Fig. 15. Model describing the shift of defect positions between the optical image and the topographical image. (**a**) Optical detection of the defect; (**b**) shear-force detection of the defect a few scan lines later

The magnetic bits are clearly visible in the magneto-optical images. Even the curvature of their edges can be identified. Lateral resolution and the magneto-optical sensitivity of the method can be determined from the contour lines along a single track (Fig. 16f): The total width of the edges amounts to about 300 nm, indicating a lateral resolution below $\lambda/2$ ($\lambda = 670$ nm); the signal-to-noise ratio is 10. Considering the known Kerr rotation of the MOD, we derive a magneto-optical sensitivity of $0.08°$ – sufficient to image magnetic domains in ultrathin films of Fe/Cu(100) [58].

We again observe a certain crosstalk between topography and optical signal. A comparison of the simultaneously obtained signals along the lines A and B reveals the cause (Fig. 16e): Contour line A exhibits a continuous drop at the groove location that is followed by a steep rise. Obviously the tip drops gradually into the groove until it is rapidly retracted again. This behavior is accompanied by an initial rise of the optical signal followed by two minima, as can be seen from the contour plot along line B. The first minimum coincides with that of line A, whereas the second one is shifted by a few hundred nm to the right with respect to the groove. We think that this behavior can be explained again by asymmetric localization of shear-force detection and light detection on the tip. The sub-λ aperture obviously passes the groove later than the topography sensor. The dips of the optical signal presumably arise from the drop of reflectivity at the precipitous groove edges.

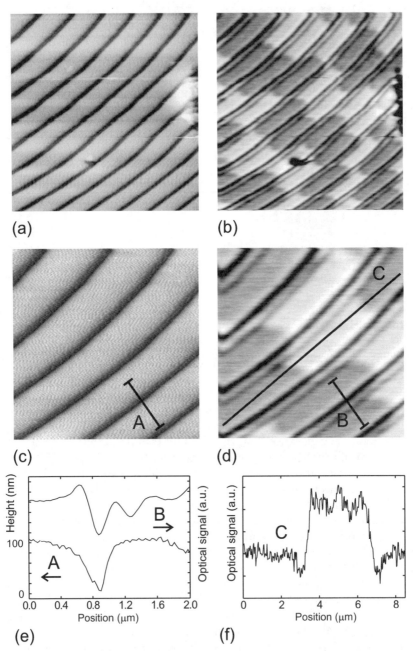

Fig. 16. $(15 \times 15)\ \mu m^2$ image of a magneto-optical disc, recorded by UHV-Sagnac-SNOM. (**a**) Topographical image; (**b**) corresponding magneto-optical image via the ω signal; (**c, d**) $(7 \times 7)\ \mu m^2$ detail images of (**a, b**) obtained by reduction of x–y scanning range; (**e, f**) contour plots along the lines A, B, and C

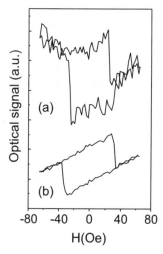

Fig. 17. Magnetization curves, locally recorded by the UHV-Sagnac-SNOM on: (a) 15-ML Ni/Cu(100) (using shear-force distance control); (b) 3.7-ML Fe/Cu(100) (using optical distance control)

Even higher magneto-optical sensitivity can be achieved using the optical distance control instead of the shear-force method. This can be seen from a comparison of locally recorded magnetization curves of ultrathin films: Figure 17a was measured on a sample of 15-ML Ni/Cu(100) with the UHV-Sagnac-SNOM using shear-force distance control. Figure 17b was measured on 3.7-ML Fe/Cu(100) (grown at 80 K) using the optical distance control. Apart from a linear background (see discussion below), both curves show a rectangular shape, as expected in polar Kerr-effect measurements for films with perpendicular easy axis of magnetization. In principle, one would expect a similar contrast for both samples because the Kerr rotation of iron is four times higher than that of Ni, and the Kerr rotation is proportional to the film thickness below the skin depth [59]. Even though the signal is slightly smaller in Fig. 17b, the signal-to-noise ratio is clearly higher.

The above-mentioned linear background in both curves is attributed to the Faraday effect caused within the glass fiber by the stray field of the electromagnet. We estimate the Faraday rotation via $\Phi_F = V\,d\,H$, where V is the Verdet constant, d the length of the glass fiber exposed to the stray field, and H the strength of the stray field. We estimate $\Phi_F \approx 12$ µrad/Oe, in agreement with the experimental observation [60].

4 Magnetic Domains in Ultrathin Films

By studying ultrathin films of Fe/Cu(100) by UHV-Sagnac-SNOM we succeeded in imaging magnetic-domain patterns, which have been observed in only a very few experiments so far. Such films exhibit a striking magnetic phenomenon, namely a spin-reorientation phase transition. In the vicinity of such a phase transition, a striped-domain phase is expected. We have studied this phase as a function of several parameters: temperature, film thickness, and external magnetic field. One question concerned the way in which stripe domains would transform into a saturated single-domain state during magnetization reversal.

4.1 Spin-Reorientation Transition

A variation of film thickness and temperature changes magnetic anisotropies. The anisotropy energy is the direction-dependent contribution to the free energy of a ferromagnet. It mainly consists of two parts: shape anisotropy, which is determined by the stray-field energy, and magnetocrystalline anisotropy, which describes preferences of magnetization orientation along particular axes of the crystal lattice [61].

Shape anisotropy results from a long-range dipolar interaction. This is the stray-field energy, which is determined by the geometry of the magnet. For thin films it is smallest for the magnetization lying in the film plane, and it has a maximum for perpendicular alignment. Shape anisotropy hence favors inplane magnetization.

The magnetocrystalline anisotropy is caused by spin-orbit coupling and crystal fields. In the bulk of cubic crystals, the angular momentum is quenched, which has the consequence that the magnetocrystalline anisotropy is usually rather small. At a surface, however, the magnetocrystalline anisotropy can be enhanced due to the reduced symmetry. The orientation-dependent part of the free energy of a thin magnetic film, G, can be written to a first approximation as:

$$G = K \cos^2 \theta = \left(K_{\text{shape}} + K_{\text{bulk}} + \frac{K_{\text{surface}}}{d} \right) \cos^2 \theta \,, \tag{9}$$

with K being the effective anisotropy constant, θ the angle between the surface normal and the magnetization vector, K_{shape} the shape-anisotropy constant ($K_{\text{shape}} = -\frac{1}{2}\mu_0 M_s^2$, M_s: saturation magnetization), K_{bulk} the 2nd-order bulk contribution to the magnetocrystalline anisotropy ($K_{\text{bulk}} = 0$ for Fe and Ni), and d the film thickness.

Opposite signs of K_{shape} and K_{surface} (e.g., in Fe) mean that these anisotropies favor orthogonal orientations. At critical values of temperature and film thickness, however, they cancel ($K = 0$) and a spin-reorientation transition (SRT) takes place (see Fig. 18). At this point, the easy axis of magnetization changes from inplane to out-of-plane. This phenomenon has been studied using MOKE in ferromagnetic ultrathin films of Fe/Ag(100) [62] and Fe/Cu(100) [58,63]. The critical film thickness was found to be in the range from 4 to 6 monolayers, depending on the substrate temperature during preparation and subsequent thermal treatment. For films prepared at 300 K

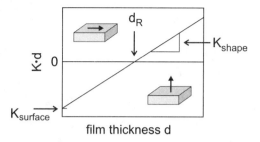

Fig. 18. Dependence of the effective anisotropy constant K on film thickness for $K_{\text{surface}} < 0$ and $K_{\text{shape}} > 0$ (e.g., in the case of Fe). A spin-reorientation transition occurs at a critical thickness d_{R}

(RT growth), the spin reorientation coincides with a phase transition of fcc Fe to bcc Fe (martensitic transition), which has not been found for films grown at 100 K (LT growth). Recent STM studies have been devoted to the crystal symmetry [64] and to the morphology [19] of such films.

4.2   Stripe-Domain Patterns

MOKE studies of the spin-reorientation phase transition of ultrathin Fe films revealed a strong reduction of the net sample magnetization in a certain interval of temperature and film thickness in the vicinity of the SRT, which in some studies was called a pseudogap [62,63]. This has been understood as a zero net magnetization due to the formation of a magnetic domain state with domain sizes far below the size of the integration area of MOKE. Such domain patterns have actually been found in ultrathin films by electron-microscopy studies [65–67]; and they consisted almost always of parallel stripes.

The experimental findings are supported by theory: The equilibrium spin configuration of a ferromagnet is a domain state [1]. Exchange interaction induces a parallel alignment of the magnetic moments over short distances (short-range order). Over larger distances, dipolar interaction favors antiparallel alignment of adjacent homogeneously magnetized domains (long-range order). This reduces stray fields, but costs energy for the formation of domain walls. The domain-wall energy of a Bloch wall is given by $4\sqrt{AK}$, with A being the exchange-stiffness coefficient. In the vicinity of a SRT, the effective anisotropy energy, K, is reduced (see. (9)) and thus the domain-wall energy is diminished, allowing for a strong reduction of the stray-field energy by formation of microdomains. In the simplest case, such a microdomain pattern consists of a periodic array of perpendicularly magnetized parallel stripes (with adjacent stripes having opposite magnetizations [68]).

Yafet and Gyorgy [69] have calculated the dependence of stripe width, a, on the anisotropy parameter, f, i.e., the ratio of the surface anisotropy energy and the dipolar energy ($f \approx 1$ at the SRT). They calculated the binding energy of the stripe-domain state, which determines the saturation

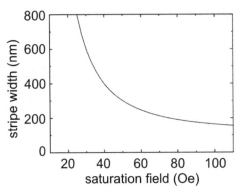

Fig. 19. Stripe width of the domain pattern in a 4-ML thick Fe film depending on the saturation field

field. It turned out that a grows exponentially with f; simultaneously, the binding energy drops. Berger and Erickson [70] extended the stripe-domain model by considering the variation of the stripe width in external magnetic fields. In this way, they calculated a phase diagram connecting f and the saturation field. Combining the information from both calculations for the case of a 4-ML thick Fe film, the stripe width can be expressed as a function of the saturation field (see Fig. 19).

The transformation of a stripe-domain pattern into the saturated, single-domain state has been studied by Kashuba and Pokrovsky [71]. They considered a stripe-domain pattern with a stripe width L, where L_0 is the equilibrium width in zero external magnetic field ($H = 0$), and δ the deviation of the stripe width from L. Calculating the energy of the pattern in an external magnetic field, they found that both δ and L diverge, if H approaches the saturation field H_S. But the width of the minority stripes, $(L - \delta)$, where the magnetization vector is antiparallel to the external magnetic field, converges to $2L_0/\pi$. This is about 60% of the equilibrium stripe width. The transformation to the saturated state can therefore be described by two parallel processes: a reduction of the minority-stripe width and an increase of the distance of minority stripes. Figure 20 shows a characteristic magnetization curve and the corresponding stripe-domain patterns.

The first work aimed at visualizing the domain-transformation process of ultrathin Fe films was recently published by Choi et al. [72]. It was a study by PEEM, and the sample was an Fe wedge under the influence of a virtual magnetic field, which was generated by interlayer exchange coupling with a Ni film via another, nonmagnetic (Cu) wedge. The effect of variable magnetic fields could be studied by moving the sample position [72]. The authors found stripe domains in the vicinity of the spin-reorientation thickness of Fe. And they observed growth of the majority-stripe width during the transformation process, with the minority stripes remaining constant in width but shrinking in length.

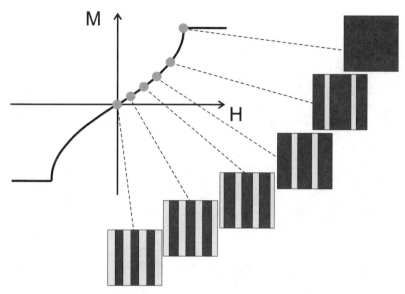

Fig. 20. Magnetization curve and corresponding transformation process of a stripe-domain pattern

Only very recently was it possible to directly observe the transformation of the stripe-domain pattern in an external magnetic field in two studies, one using spin-polarized low-energy electron microscopy (SPLEEM) [67] and the other one using the present UHV-Sagnac-SNOM setup.

4.3 Domain Contrast

We first had to demonstrate that UHV-Sagnac-SNOM is actually able to image magnetic domains in ultrathin films. We checked this for a 3.7-ML thick film of Fe/Cu(100), well below the SRT, where rather large domains with perpendicular magnetization should be favored. A 550×550 μm^2 Kerr-microscopy image of this film is shown in Fig. 21. The random domain pattern appeared upon demagnetization of the film. Using the UHV-Sagnac-SNOM, it is possible to study this pattern in detail: 11×20 μm^2 sized scans of the domain structure are shown in Fig. 22a–c. These images have been obtained by successively moving the scan position a few micrometers to the right from image to image.

These are the first domain images of an ultrathin film, obtained in situ by SNOM. We now discuss the magnetic contrast by comparing the topographical and magneto-optical images obtained from the 3.7-ML film of Fe/Cu(100) shown in Fig. 23. The topographical image corresponding to the magneto-optical image given in Fig. 23b is shown in Fig. 23a. As the optical distance control is used here, this image represents the contour of a constant 2ω signal.

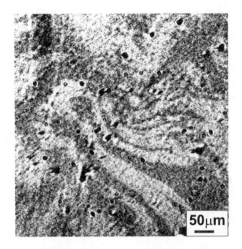

Fig. 21. Kerr-microscopy image of the domain pattern in a demagnetized 3.7-ML thick film of Fe/Cu(100), grown at 80 K, annealed at 345 K, and measured at 285 K

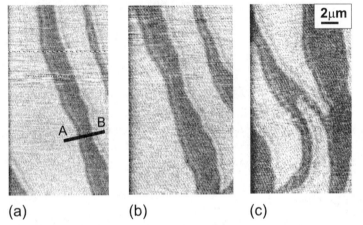

(a)   (b)   (c)

Fig. 22. UHV-Sagnac-SNOM images of magnetic domains in a demagnetized 3.7-ML-thick film of Fe/Cu(100). The scan position was moved by a few micrometers to the right from image to image

The domain structures are not visible in the topography. Thus, magnetic and nonmagnetic signals can obviously be distinguished using the Sagnac interferometer. The only exception is a semicircular defect in the topography, which gives rise to some crosstalk. The pure magnetic origin of the domain pattern can be clearly seen if the same sample position is scanned twice with a short magnetic field pulse being applied after 40% of the second scan, which transforms the domain state into a magnetically saturated single-domain state. As can be seen in Fig. 23c, the domain pattern is erased by the magnetic-field pulse. Only the crosstalk structure remains.

Figure 24a represents the contour of the optical signal along the line AB in Fig. 22a. From this, we conclude that the width of the domain walls amounts

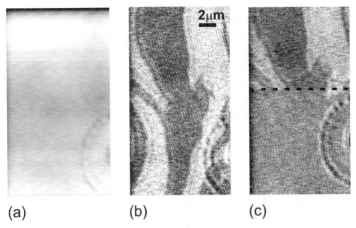

(a)   (b)   (c)

Fig. 23. UHV-Sagnac-SNOM images of magnetic domains in a 3.7-ML film of Fe/Cu(100): (a) topographical and (b) corresponding magneto-optical signal; (c) saturation of the domain pattern upon application of a 150-Oe magnetic-field pulse at the scan position indicated by a *dashed line*. The semicircular defect comes from crosstalk between topography and magnetic signal

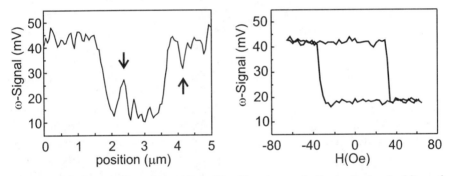

Fig. 24. (a) Contour line along AB in Fig. 22a. *Arrows* indicate dip by doubling of the domain-wall shape. (b) Locally recorded magnetization curve of the same film. The curve is that from Fig. 17b, however, after subtraction of a linear background due to the Faraday effect

to 300 nm. Hence, lateral resolution is below $\lambda/2$ here as well; the signal-to-noise ratio is 6:1. From the known Kerr rotation of Fe/Cu(100) at this particular thickness, we derive a magneto-optical sensitivity of 0.02°. The magneto-optical contrast of oppositely magnetized adjacent domains is comparable to the difference of the signals corresponding to the opposite saturation states of a locally recorded magnetization curve (compare Figs. 24a–b).

One SNOM artifact present in all images is the double structure of the domain walls. Moved by a few hundred nanometers to the right, the shape of the wall between light and dark domains appears again as a narrow line.

This can be seen in the contour plot, as well (Fig. 24a). We can exclude phase contrasts or relative delay in processing of signals. Therefore, we think that this behavior is caused by some kind of double-tip effect. A small metal particle, picked up by the glass-fiber tip, e.g., might act as a second near-field probe, being slightly moved relative to the symmetry axis of the tip. It is not yet clear whether the second wall is a reduction of total intensity or a magnetic signal. A magnetic signal would mean that the second tip detects the components of the domain-wall magnetization lying in the plane, which would be a pure near-field Kerr effect, similar to the one observed by Silva et al. [38]. Further experiments are needed to clarify the origin of this effect.

4.4   Study of Magnetization Reversal

The combination of all three magneto-optical imaging methods that are available in our experimental setup was used to study the domain transformation upon magnetization reversal of ultrathin films. By polar MOKE we recorded a series of magnetization curves of 4.2-ML Fe/Cu(100) in the vicinity of the SRT (see Fig. 25). For ultrathin Fe films, the SRT manifests itself as a transition of polar-MOKE curves from a rectangular easy-axis shape at low temperatures ($K > 0$) to a hard-axis shape at high temperatures ($K < 0$). This can clearly be seen in Fig. 25, where we find full remanence at 260 K and an almost flat shape at 300 K. In between, the curves exhibit a shape comparable to the one predicted by theory for the vicinity of the SRT (see Fig. 20). In contrast to theory, however, the curves flatten at the saturation field and show strong hysteresis effects if the temperature is lowered. The inner loop at 280 K clearly shows that the saturated state is only metastable on reduction of the external magnetic field. Assuming that the ground-state configuration is a domain state with zero remanence, we can conclude already at this point that the transformation processes between the stripe domains and the single-domain state will require thermal activation to compensate for pinning of domains at local inhomogeneities of the sample.

It should be noted here that magnetization curves like the one measured at 280 K have been found at the SRT of several other systems in the past, e.g., Fe/Cu$_3$Au [73] and Fe/Ag(100) [74].

Looking for domain patterns at the SRT of 4.2-ML Fe/Cu(100), we first applied Kerr microscopy after adjusting the temperature to get a polar-MOKE curve with an almost vanishing coercive field H_c. The Kerr-microscopy images show the formation of a (dark) domain at H_c: The metastable single-domain state (see Fig. 26a) is converted almost suddenly by domain nucleation and domain-wall displacement into that new state (see Fig. 26b). According to the magnetization curve, this state is not just oppositely magnetized. It rather has a reduced net magnetization, which gradually switches into the oppositely saturated state on raising the magnetic field. If the external magnetic field is reduced, the domain pattern remains unchanged, but the contrast decreases. However, it still remains visible, even in a reversed

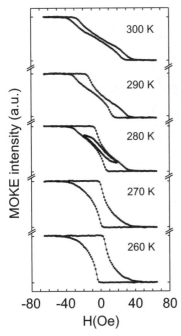

Fig. 25. Temperature dependence of the shape of magnetization curves at the spin-reorientation transition of 4.2-ML Fe/Cu(100) obtained by polar MOKE. (The film was grown at 80 K and annealed at 300 K)

magnetic field. At this point, the variation of contrast can be explained in two ways: (i) The new state favors an inplane magnetization, which rotates out of the plane when the perpendicular magnetic field is increased; (ii) it is a pattern of perpendicularly magnetized microdomains that transform into a saturated single-domain state by domain growth and nucleation, but remain unresolved for far-field Kerr microscopy.

UHV-Sagnac-SNOM allowed us finally to determine the inner structure of the phase that nucleates at H_c. The images and plots presented in Fig. 27 were recorded on the same sample as those given in Fig. 26. The magneto-optical image in Fig. 27a clearly shows that the area with a reduced net magnetization consists actually of striped microdomains. By comparing contour plots we found the light stripes in the image to have more or less the same widths, namely (380 ± 30) nm. This is far below the resolving power of our UHV-Kerr microscope but can be resolved with our UHV-Sagnac-SNOM. We assume that the stripes already have their equilibrium width, even though there is a strong imbalance of light stripes and dark surroundings. This remanence at $H = 0$ could already be seen in the magnetization curves (see Fig. 25). It results from the irreversibility of the growth process induced by pinning effects. Obviously, the stripes are not able to move as freely as as-

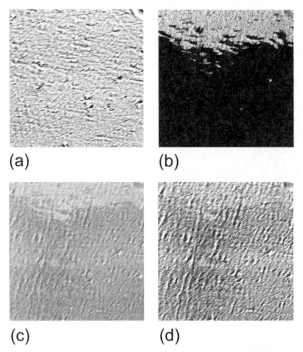

Fig. 26. 550 μm × 550 μm Kerr-microscopy images of 4.2-ML Fe/Cu(100) recorded at a sample temperature of 150 K in variable external magnetic fields: (**a**) homogeneously magnetized film; (**b**) during domain growth at the coercive field (H_c); (**c**) at reduced fields ($0 < H < H_c$); (**d**) at reversed fields ($-H_c < H < 0$)

sumed in the theoretical considerations mentioned above. They merely grow in length and in a two-dimensional way (see, e.g., the turnarounds of stripes) by thermally activated steps (Barkhausen jumps). However, in some parts of the image, they are rather dense, as if pinning were less strong there. Nevertheless, the stripes seem to favor parallel alignment, which indicates strong orientational order but reduced positional order. Therefore, stripe domains have sometimes been described in analogy to liquid crystals [75,66].

The measured stripe widths can be compared to theoretically expected values: From the polar-MOKE hysteresis curve, we obtain a saturation field $H_0 = (45 \pm 5)$ Oe, which, according to Fig. 19, translates into a stripe width of (345 ± 50) nm. This is in excellent agreement with the measured value of (380 ± 30) nm. The connection of stripe widths and saturation field (Fig. 19) could thus be confirmed by this experiment.

The assumption of a thermally activated magnetization-reversal process is supported by the shape of the magnetization curves (see Fig. 25): Reduction of sample temperature increases the saturation field. A purely reversible process would behave completely different: Crystalline anisotropy increases

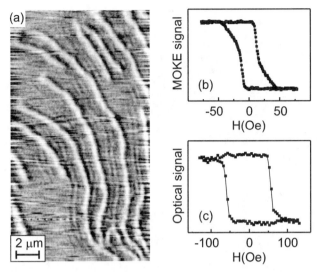

Fig. 27. Stripe-domain pattern observed on 4.2-ML Fe/Cu(100) by UHV-Sagnac-SNOM at a sample temperature of 200 K. (**a**) Domain state in remanence, after a reversed field of 30 Oe had been applied. (**b**) Polar-MOKE curve of the film, indicating the vicinity of a spin-reorientation transition, (**c**) local hysteresis curve recorded by UHV-Sagnac-SNOM at a fixed sample position (after subtraction of a linear background due to the Faraday effect.) This represents the Barkhausen jump of a domain wall

when the sample temperature is reduced. This would reduce the binding energy of a domain state, and hence, also the saturation field. Such behavior can be seen in magnetization curves at higher temperatures (290 K, 300 K) as well, where pinning effects are weak.

4.5 Transformation of Stripe Domains

We have studied the transformation of stripe domains by imaging the pattern at variable external magnetic fields with the UHV-Sagnac-SNOM. This is shown for the 4.2-ML thick film in Fig. 28. The same sample position was successively scanned at −15 Oe, 0 Oe, and +15 Oe (see Fig. 28a–c). Figure 28d was obtained on a slightly different position of the sample in a magnetic field of +30 Oe, which is still below saturation according to the magnetization curve (see Fig. 27b). By looking at the series of images, we observe the following general features: Stripes with magnetization parallel to the magnetic field (majority stripes) are stabilized, whereas those with antiparallel orientation (minority stripes) become unstable. Obviously, the minority stripes do not change their widths, and they are pinned to certain sample positions, where inhomogeneities of the surface (e.g., from polishing scratches) affect the anisotropies. Such pinning induces a certain deviation of the growth pro-

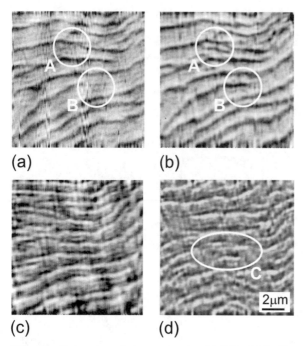

Fig. 28. Stripe-domain pattern of a 4.2-ML thick film, recorded by UHV-Sagnac-SNOM at various external magnetic fields: (**a**) at −15 Oe, (**b**) at 0 Oe, (**c**) at +15 Oe, (**d**) at +30 Oe. Comparing the images, nucleation (A) and length growth (B) of domains can be detected, as well as growth in width (compare (**b**) and (**c**)). Close to saturation, a droplet phase occurs (**d**). The images are slightly Fourier filtered, scan direction was vertical, and (**d**) was obtained at a different position on the sample from (**a–c**)

cess from the theoretically expected behavior (see Fig. 20). Instead of moving closer together on decreasing external magnetic field, the minority stripes grow in length or new stripes nucleate to fill the domain pattern more densely (see areas A and B in Fig. 28). The transformation of the domain pattern upon reversal of the external magnetic field is also governed by pinning effects. It looks like a contrast reversal (compare Figs. 28b–c), but it is actually just a widening of the minority stripes and a narrowing of majority stripes (which then turn into minority stripes). Another interesting phenomenon can be seen close to the saturation field (see Fig. 28d): The minority stripes break up into short pieces. Such a transition to a droplet phase has recently also been observed for the same system by SPLEEM [67]. In principle, this is well understood: In the vicinity of the saturation field, the distance between minority stripe domains is relatively large. This reduces the magnetostatic interaction between the stripes. Concentration of parallel magnetic moments to a few narrow stripes is thermally unfavorable. If less than 30% of the area is covered by stripe domains, a state consisting of droplet-like domains has lower energy than stripes. But since the energy gain is less than 10% [76] spontaneous transitions between stripe and droplet domains occur rarely. Pinning of stripe domains could help here to open a channel for such a transition.

Inhomogeneities are almost unavoidable in ultrathin films. Pinning of stripe domains has therefore been reported by other groups as well: Choi et al. [72] claim that their domain images of Fe/Cu(100) obtained by PEEM in a virtual magnetic field can be explained if pinning is assumed. The images of the droplet phase, obtained by SPLEEM [67] obviously show pinning. Also, recent SEMPA studies of the thermally induced transformation processes of stripe domains in ultrathin films of Fe/Cu(100) show a certain positional order in addition to the orientational order of stripes, which can be attributed to pinning [77].

The quintessence of our studies so far is a model of the magnetization-reversal process of ultrathin films in the vicinity of a spin-reorientation phase transition. Taking into account pinning effects, we suggest a two-dimensional growth process via defect-induced branching and domain nucleation. Our model is summarized in Fig. 29: Curve A represents the shape of the magnetization curve of a film without pinning, as is predicted by theory (see Fig. 20). In addition, we show a curve (C) as is usually observed experimentally (see Fig. 25). Lower temperatures cause stronger pinning with two effects on the shape of the curves: (i) A larger coercive field since thermal fluctuations are smaller and larger magnetic fields are necessary to overcome activation barriers for nucleation and growth of stripe domains. (ii) An increased saturation field for the same reason. The transformation of the stripe-domain pattern in a variable external magnetic field can be observed best for an inner loop (curve B in Fig. 29) where the maximum applied magnetic field does not exceed the saturation field. Here, we do not have to take into account the generation process of the stripe-domain phase out of the saturated single-domain state, which would add another degree of complexity.

Starting with small droplet-shaped domains (1) just below the saturation field, branching of domains occurs upon reduction of the external magnetic field, because the minority domains require more space (2). Locations with enhanced anisotropy, like surface defects, play the role of branching points. Locations with lowered anisotropy, on the other hand, act as nucleation centers for new domains. Polishing scratches, e.g., support the formation of stripe domains in this way. We believe that very few polishing scratches can determine the domain pattern in a wide area due to the orientational order of stripe domains (3). On reversal of the external magnetic field, the stripe domains grow in width (4), but due to pinning, they do not change their positions. Minority stripes turn into majority stripes and vice versa. Further increase of the external magnetic field induces a shortening of the stripes (5). Stripes that are not pinned may drift apart, whereas pinned stripes break up and form droplets. This generates a balanced distribution of minority domains at low coverages in the vicinity of the saturation field (6).

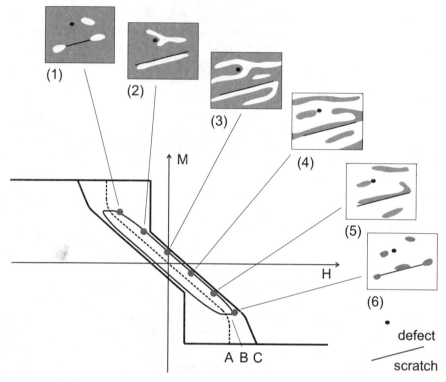

Fig. 29. Model of the magnetization-reversal process in ultrathin films in the vicinity of a spin-reorientation transition taking into account pinning effects. Magnetization curves: (A) without pinning, (C) with pinning, (B) inner hysteresis loop of C. (1)–(6) Transformation of the stripe-domain pattern in the magnetization-reversal process

5   Summary and Future Prospects

By using an UHV-Sagnac-SNOM, we succeeded in imaging the stripe-domain patterns of ultrathin films of Fe/Cu(100), which is the first application of magneto-optical SNOM to samples other than test patterns. Stripe domains are the ground-state spin configuration of ultrathin films in the vicinity of a spin-reorientation transition. It was not even clear at the beginning whether the stripe-domain state would play any role during the magnetization-reversal process. Previous data could also be explained by a coherent rotation of the magnetization vector within a single-domain state. The observation of stripe domains in external magnetic fields by UHV-Sagnac-SNOM shows unambiguously that the magnetization-reversal process takes place by transformation of a domain phase. MOKE and Kerr-microscopy data can be explained in this way as well. As expected, the stripe width depends on the saturation field of the corresponding magnetization curve. It turned out that pinning of stripe domains plays a major role during magnetization reversal of ultra-thin films. Surface defects, like polishing scratches, obviously determine the orientational order of the domain stripes and induce a positional order that prevents the stripes from drifting apart. Close to saturation, the stripes break up into droplets.

We believe that the combination of MOKE, Kerr microscopy and UHV-Sagnac-SNOM is a very powerful approach for studying thin magnetic films. We demonstrated that SNOM works well in UHV and that a Sagnac interferometer is a more appropriate method for Kerr-effect detection than a conventional crossed-polarizer setup. We also found that the total-intensity signal provided by the Sagnac interferometer can effectively be used for distance control, which in UHV is superior to the common shear-force method. SNOM has the advantage that it can be operated in external magnetic fields. The present setup provides a lateral resolution better than $\lambda/2$ and a magneto-optical sensitivity of better than $0.02°$. This can be derived from domain images of ultrathin Fe films. It will be difficult to achieve convincing magnetic-domain contrast with lateral resolution better than 100 nm using uncoated glass-fiber tips. But the progress in apertureless SNOM might soon provide a reasonable way of probing the Kerr effect.

We expect another aspect of magneto-optics to play a major role in future SNOM experiments: Its ability to resolve ultrafast magnetization dynamics. This has been demonstrated successfully in far-field pump-probe experiments so far. Either an all-optical setup was combined with Kerr-effect detection [78,79], or a short magnetic-field pulse, synchronized with the Kerr-effect-probing laser, was used to excite the dynamical states [5,80,81]. For laterally resolved studies, such stroboscopic experiments have to be performed at each scan position. In this way, even time-resolved magneto-optical SNOM should be feasible. Time resolution in the femtosecond range has recently been achieved by SNOM, e.g., in the study of the nonlinear optical response of single semiconductor quantum dots [82]. For time-resolved magneto-optical studies it might again be advantageous to use a Sagnac interferometer: If the pulse of a fs-laser source were split into counterpropagating pulses of different intensity, one of them could be used as the pump, and the other one could act as the probe pulse. Time delay between both could be created by insertion of delay stages into the Sagnac interferometer. The nonreciprocal phase difference would be a measure of the transient changes of the magnetization between pump and probe pulses.

Ultrathin films would be interesting samples for time-resolved studies. In the vicinity of the spin-reorientation transition, an increase of temperature could induce spin precession because it changes the easy axis of the magnetization. The heat pulse acts like an effective perpendicular magnetic field pulse, which according to the Landau–Lifshitz equation excites spin precession [83]. Such thermally induced magnetization dynamics has already been studied by time-resolved MOKE [79].

38   G. Meyer, A. Bauer, and G. Kaindl

Acknowledgements

The authors gratefully acknowledge assistance and support by T. Crecelius, I. Mauch, and D. Wegner (all from Freie Universität Berlin, Germany). This work was financially supported by the Deutsche Forschungsgemeinschaft, SfB 290, TPA06.

References

1. A. Hubert, R. Schäfer: *Magnetic Domains* (Springer, Berlin 1998)
2. M. Bode: Rep. Prog. Phys. **66**, 523 (2003)
3. H.J. Hug, B. Stiefel, P.J.A. van Schendel, A. Moser, R. Hofer, S. Martin, H.-J. Güntherodt, S. Porthun, L. Abelmann, J.C. Lodder, G. Bochi, R.C. O'Handley: J. Appl. Phys. **83**, 5609 (1998)
4. J. Unguris: 'Scanning Electron Microscopy with Polarization Analysis'. In: *Magnetic Microscopy and its Applications to Materials.* ed. by M. De Graef (Academic Press, San Diego 2000)
5. B.C. Choi, M. Belov, W.K. Hiebert, G.E. Ballentine, M.R. Freeman: Phys. Rev. Lett. **86**, 728 (2001)
6. T. Eimüller, P. Fischer, M. Köhler, M. Scholz, P. Guttmann, G. Denbeaux, S. Glück, G. Bayreuther, G. Schmahl, D. Attwood, G. Schütz: Appl. Phys. A **73**, 697 (2001)
7. J. Stöhr, Y. Wu, B.D. Hermsmeier, M.G. Samant, G.R. Harp, S. Koranda, D. Dunham, B.T. Tonner: Science **259**, 658 (1993)
8. B.L. Petersen, A. Bauer, G. Meyer, T. Crecelius, G. Kaindl: Appl. Phys. Lett. **73**, 538 (1998)
9. A. Bauer, B.L. Petersen, T. Crecelius, G. Meyer, D. Wegner, G. Kaindl: J. Microscopy **194**, 507 (1999)
10. G. Meyer, T. Crecelius, G. Kaindl, A. Bauer: J. Magn. Magn. Mater. **240**, 76 (2002)
11. G. Meyer, T. Crecelius, A. Bauer, D. Wegner, I. Mauch, G. Kaindl: J. Microscopy **210**, 209 (2003)
12. G. Meyer, A. Bauer, T. Crecelius, I. Mauch, G. Kaindl: Phys. Rev. B **68**, 212404 (2003)
13. G. Meyer, T. Crecelius, A. Bauer, I. Mauch, G. Kaindl: Appl. Phys. Lett. **83**, 1394 (2003)
14. A.K. Zvezdin, V.A. Kotov: *Modern Magnetooptics and Magnetooptical Materials.* (Institute of Physics Publishing, Bristol 1997)
15. P.M. Oppeneer, T. Maurer, J. Sticht, J. Kubler: Phys. Rev. B **45**, 10924 (1992)
16. J. Schoenes: 'Magneto-Optical Properties of Metals, Alloys and Compounds'. In: *Materials Science and Technology, Vol. 3A Electronic and Magnetic Properties of Metals and Ceramics Part 1.* ed. by K.H.J. Buschow (VCH, Weinheim 1991)
17. S. Ezekiel, H.J. Arditty: *Fiber-Optic Rotation Sensors and Related Technologies.* (Springer, Berlin 1982)
18. S. Spielman: Optical tests for broken time-reversal symmetry in the cuprate superconductors, Ph.D.-Thesis, Ginzton Report No. 4961, Stanford University (1992)

19. E. Mentz, D. Weiss, J.E. Ortega, A. Bauer, G. Kaindl: J. Appl. Phys. **82**, 482 (1997)
20. D. Peterka, A. Enders, G. Haas, K. Kern: Rev. Sci. Instrum. **74**, 2744 (2003)
21. A. Vaterlaus, U. Maier, U. Ramsperger, A. Hensch, D. Pescia: Rev. Sci. Instrum. **68**, 2800 (1997)
22. D.W. Pohl, W. Denk, M. Lanz: Appl. Phys. Lett. **44**, 651 (1984)
23. E. Betzig, J.K. Trautmann, T.D. Harris, J.S. Weiner, R.L. Kostelak: Science **251**, 1468 (1991)
24. M.A. Paesler, P.J. Moyer: *Near-Field Optics*. (John Wiley, New York 1996)
25. V. Sandoghdar, S. Wegscheider, G. Krausch, J. Mlynek: J. Appl. Phys. **81**, 2499 (1997)
26. A. Rosenberger, A. Münnemann, F. Kiendl, G. Güntherodt, P. Rosenbusch, J.A.C. Bland, G. Eggers, P. Fumagalli: J. Appl. Phys. **89**, 7727 (2001)
27. E. Betzig, J.K. Trautman, R. Wolfe, E.M. Gyorgy, P.L. Finn, M.H. Kryder, C.H. Chang: Appl. Phys. Lett. **61**, 142 (1992)
28. T. Lacoste, T. Huser, H. Heinzelmann: Z. Phys. B **104**, 183 (1997)
29. G. Eggers, A. Rosenberger, N. Held, P. Fumagalli: Surf. Interface Anal. **25**, 483 (1997)
30. V. Kottler, N. Essaidi, N. Ronarch, C. Chappert, Y. Chen: J. Magn. Magn. Mater. **165**, 398 (1997)
31. Y. Mitsuoka, K. Nakajima, K. Homma, N. Chiba, H. Muramatsu, T. Ataka, K. Sato: J. Appl. Phys. **83**, 3998 (1998)
32. P. Bertrand, L. Conin, C. Hermann, G. Lampel, J. Peretti: J. Appl. Phys. **83**, 6834 (1998)
33. O. Bergossi, H. Wioland, S. Hudlet, R. Deturche, P. Royer: Jpn. J. Appl. Phys. **38**, 655 (1999)
34. H. Wioland, O. Bergossi, S. Hudlet, K. Mackay, P. Royer: Eur. Phys. J. AP **5**, 289 (1999)
35. W. Dickson, S. Takahashi, R. Pollard, R. Atkinson, A.V. Zayats: J. Microscopy **209**, 194 (2003)
36. C. Durkan, I.V. Shvets, J.C. Lodder: Appl. Phys. Lett. **70**, 1323 (1997)
37. P. Fumagalli, A. Rosenberger, G. Eggers, A. Münnemann, N. Held, G. Güntherodt: Appl. Phys. Lett. **72**, 2803 (1998)
38. T.J. Silva, S. Schultz, D. Weller: Appl. Phys. Lett. **65**, 658 (1994)
39. L. Aigouy, S. Grésillon, L. Lahrech, A.C. Boccara, J.C. Rivoal, V. Mathet, C. Chappert, J.P. Jamet, J. Ferré: J. Microscopy **194**, 295 (1999)
40. T.J. Silva, S. Schultz: Rev. Sci. Instrum. **67**, 715 (1996)
41. G. Eggers, A. Rosenberger, N. Held, G. Güntherodt, P. Fumagalli: Appl. Phys. Lett. **79**, 3929 (2001)
42. C. Durkan, I.V. Shvets: J. Appl. Phys. **83**, 1171 (1998)
43. V.A. Kosobukin: Tech. Phys. **43**, 824 (1998)
44. J.N. Walford, J.A. Porto, R. Carminati, J.J. Greffet: J. Opt. Soc. Am. A **19**, 572 (2002)
45. R. Müller, C. Lienau: Appl. Phys. Lett. **76**, 3367 (2000)
46. A. Kapitulnik, J.S. Dodge, M.M. Fejer: J. Appl. Phys. **75**, 6872 (1994)
47. D.R. Turner: U.S. Patent No. 4,469,554 (1984)
48. R. Stöckle, C. Fokas, V. Deckert, R. Zenobi, B. Sick, B. Hecht, U.P. Wild: Appl. Phys. Lett. **75**, 160 (1999)
49. M. Ohtsu, K. Sawada: 'High-Resolution and High-Throughput Probes'. In: *Nano-Optics*. ed. by S. Kawata, M. Ohtsu, M. Irie (Springer, Berlin 2002)

50. S.I. Bozhevolnyi, B. Vohnsen: J. Opt. Soc. Am. B **14**, 1656 (1997)
51. W.A. Atia, S. Pilevear, A. Güngör, C.C. Davis: Ultramicroscopy **71**, 379 (1998)
52. R. Brunner, A. Bietsch, O. Hollricher, O. Marti: Rev. Sci. Instrum. **68**, 1769 (1997)
53. P. Anger, A. Feltz, T. Berghaus, A.J. Meixner: J. Microscopy **209**, 162 (2003)
54. R. Brunner, O. Marti, O. Hollricher: J. Appl. Phys. **86**, 7100 (1999)
55. M.J. Gregor, P.G. Blome, J. Schöfer, R.G. Ulbrich: Appl. Phys. Lett. **68**, 307 (1996)
56. K. Karrai, I. Tiemann: Phys. Rev. B **62**, 13174 (2000)
57. M. Mansuripur: *The Physical Principles of Magneto-optical Recording*. (Cambridge University Press, Cambridge 1995)
58. S. Müller, P. Bayer, C. Reischl, K. Heinz, B. Feldmann, H. Zillgen, M. Wuttig: Phys. Rev. Lett. **74**, 765 (1995)
59. S.D. Bader: J. Magn. Magn. Mater. **100**, 440 (1991)
60. G. Meyer: In situ Abbildung magnetischer Domänen in dünnen Filmen mit magnetooptischer Rasternahfeldmikroskopie. Dissertation, Freie Universität Berlin (2003)
61. W.J.M. de Jonge, P.J.H. Bloemen, F.J.A. den Broeder: 'Experimental Investigations of Magnetic Anisotropy'. In: *Ultrathin Magnetic Structures I*. ed. by J.A.C. Bland, B. Heinrich (Springer, Berlin 1994)
62. Z.Q. Qiu, J. Pearson, S.D. Bader: Phys. Rev. Lett. **70**, 1006 (1993)
63. D.P. Pappas, K.P. Kämper, H. Hopster: Phys. Rev. Lett. **64**, 3179 (1990)
64. A. Biedermann, R. Tscheließnig, M. Schmid, P. Varga: Phys. Rev. Lett. **87**, 086103 (2001)
65. R. Allenspach, A. Bischof: Phys. Rev. Lett. **69**, 3385 (1992)
66. A. Vaterlaus, C. Stamm, U. Maier, M.G. Pini, P. Politi, D. Pescia: Phys. Rev. Lett. **84**, 2247 (2000)
67. R.J. Phaneuf, A.K. Schmid: Physics Today **56**(3), 50 (2003)
68. P. Jensen: Magnetische Eigenschaften dünner ferromagnetischer Filme. Habilitation Thesis, Freie Universität Berlin, 1994
69. Y. Yafet, E.M. Gyorgy: Phys. Rev. B. **38**, 9145 (1988)
70. A. Berger, R.P. Erickson: J. Magn. Magn. Mater. **165**, 70 (1997)
71. A. Kashuba, V.L. Pokrovsky: Phys. Rev. B **48**, 10335 (1993)
72. H.J. Choi, W.L. Ling, A. Scholl, J.H. Wolfe, U. Bovensiepen, F. Toyama, Z.Q. Qiu: Phys. Rev. B **66**, 014409 (2002)
73. F. Baudelet, M.T. Lin, W. Kuch, K. Meinel, B. Choi, C.M. Schneider, J. Kirschner: Phys. Rev. B **51**, 12563 (1995)
74. A. Berger, H. Hopster: Phys. Rev. Lett. **76**, 519 (1996)
75. A. Abanov, V. Kalatsky, V.L. Pokrovsky, W.M. Saslow: Phys. Rev. B **51**, 1023 (1995)
76. K.O. Ng, D. Vanderbilt: Phys. Rev. B **52**, 2177 (1995)
77. O. Portmann, A. Vaterlaus, D. Pescia: Nature **422**, 701 (2003)
78. E. Beaurepaire, J.C. Merle, A. Daunois, J.Y. Bigot: Phys. Rev. Lett. **76**, 4250 (1996)
79. B. Koopmans, M. van Kampen, J.T. Kohlhepp, W.J.M. de Jonge: Phys. Rev. Lett. **85**, 844 (2000)
80. Y. Acremann, C.H. Back, M. Buess, O. Portmann, A. Vaterlaus, D. Pescia, H. Melchior: Science **290**, 492 (2000)
81. J.P. Park, P. Eames, D.M. Engebretson, J. Berezovsky, P.A. Crowell: Phys. Rev. Lett. **89**, 277201 (2002)

82. T. Günther, C. Lienau, T. Elsässer, M. Glanemann, V.M. Axt, T. Kuhn, S. Eshlaghi, A.D. Wieck: Phys. Rev. Lett. **89**, 057401 (2002)
83. R.L. Stamps, B. Hillebrands: Appl. Phys. Lett. **75**, 1143 (1999)

Improvement of Interface Quality in Cleaved-Edge-Overgrowth GaAs Quantum Wires Based on Micro-optical Characterization

Masahiro Yoshita and Hidefumi Akiyama

1 Introduction

Low-dimensional semiconductor nanostructures such as quantum wires and dots have attracted great attention both in fundamental physics and in device applications due to their novel properties inherent to low dimensionality. For the realization of novel semiconductor nanostructures with high spatial uniformity, development of the fabrication methods and characterization of the nanostructures have been extensively performed in recent decades [1,2].

The quantum wire is a promising candidate for highly functional next-generation electronic and optoelectronic devices. Also, from the viewpoint of fundamental physics, due to strong Coulomb interactions, the appearance of novel one-dimensional (1D) physical phenomena is expected. The semiconductor quantum wires are fabricated by advanced epitaxial crystal growth techniques such as molecular beam epitaxy (MBE) and metalorganic chemical vapor deposition (MOCVD). Recent progress in these growth techniques has realized various kinds of quantum-wire structures, in which novel phenomena such as enhancement of exciton binding energy, concentrated oscillator strength of 1D excitons, squeezing of the wave functions of 1D excitons, high-density excitation effects, and lasing have been studied [3–22].

However, in these quantum-wire structures, relatively large structural inhomogeneity due to heterointerface roughness still exists because the epitaxial growth of the quantum wire structures is difficult, in contrast to that of the quantum wells (QWs) where the epitaxial growth is two-dimensional (2D) on the flat surfaces. This structural inhomogeneity causes localization of the electronic states of the quantum wire into zero-dimensional quantum dots (QDs) at a low temperature, and disturbs inherent 1D properties [23–25]. Complete physical understanding of the observed phenomena such as exciton Mott transition to an electron–hole plasma or lasing of quantum-wire lasers still remains controversial. Therefore, realization of high-quality semiconductor quantum wires showing novel 1D properties becomes important and urgent. To reach the goal of this challenging issue, microscopic evaluation of the electronic, optical, and structural properties of the quantum wires that returns essential feedback for further development of a novel epitaxial growth technique for reducing structural inhomogeneity is quite important.

In this chapter, we aim at describing how we have achieved high-quality T-shaped quantum wires (T wires) with high spatially uniformity by using a cleaved-edge overgrowth (CEO) method with MBE. To characterize local structural and optical properties in T wires, we here use a high-resolution microscopic photoluminescence (micro-PL) method as a local probe technique [26]. We first, in Sects. 2 and 3, examine the origins of the structural inhomogeneity existing in the T wires grown by the original CEO method by using micro-PL imaging and spectroscopy. From micro-PL spectroscopy, we find that the large interface roughness exists in the (110) epitaxial layer grown by the CEO method, and that causes structural inhomogeneity of the T wires. To reduce the surface roughness existing in the (110) GaAs CEO layer, in Sect. 4 we develop a modified CEO method combined with a growth-interrupt in situ annealing technique. By using this technique, the surface roughness is dramatically reduced and an atomically flat surface over several tens of μm in extent is formed on the CEO surface. Moreover, on the basis of the obtained surface morphology by means of atomic force microscopy (AFM), flat-surface formation mechanisms on the (110) GaAs during growth-interrupt annealing are also discussed. In Sect. 5 we fabricate a (110) GaAs QW with atomically smooth interfaces by using the modified CEO method combined with the growth-interrupt annealing, and investigate the interface properties of the QW by means of micro-PL spectroscopy and imaging. From the micro-PL results, we confirm the effectiveness of the modified CEO method on the growth of high-quality CEO QW. Finally, in Sect. 6, we realize a high-quality T wire by using the modified CEO method, and characterize spatial uniformity of the wire states by the micro-PL technique. We also demonstrate lasing from a single-T-wire laser structure with optical pumping.

2   T-Shaped Quantum Wires Grown by Cleaved-Edge Overgrowth Method

2.1   Cleaved-Edge Overgrowth Method with MBE

The T wires are fabricated by the CEO method, in which two MBE growth steps are separated by an in situ wafer-cleavage process [27]. The procedure of the CEO method is schematically shown in Fig. 1. In the first growth step, QWs are grown, by the conventional MBE growth, on a (001) substrate. After the first growth, the (001) substrates are taken out from the MBE chamber, thinned from the backside to around 80–100 μm in thickness, and partly scribed to have incipient cleavage. The thinned substrates are mounted vertically on the sample holder, and reloaded into the MBE chamber. The in situ cleavage is carried out to expose a fresh (110) edge surface. Then, the second MBE growth is performed on the (110) edge formed by the in situ cleavage. A typical T-wire structure is schematically shown in Fig. 2. Quantum-wire electronic states are quantum-mechanically confined at a T-intersection of

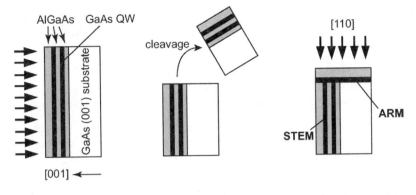

1. 1st MBE growth   2. In situ cleavage   3. 2nd MBE growth

Fig. 1. Schematics of the cleaved-edge overgrowth method with molecular beam epitaxy

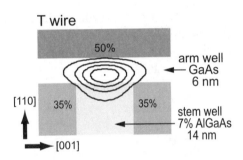

Fig. 2. Schematic of a T-shaped quantum-wire structure that consists of a 14-nm thick (001) QW (stem well) and a 6-nm thick (110) QW (arm well). Percentages show Al contents (x) in Al$_x$Ga$_{1-x}$As layers. The *contour curves* show constant probability ($|\psi|^2 = 0.2$, 0.4, 0.6, 0.8, and 1.0) for electrons confined in the quantum-wire structure

the (001) QW in the first MBE growth, which is denoted as a stem well, and the (110) QW overgrown on the (110) edge, which is denoted as an arm well.

The CEO method with MBE has been used to fabricate a host of low-dimensional quantum structures including T wires [3–9,16,17,28–31], modulation-doped quantum wires showing nearly ideal quantum transport characteristics [32], and precisely spaced QDs [33].

The advantages of the CEO method for fabricating quantum wires are (1) precise control of layer thickness with an atomic scale in the QWs because each MBE growth is a 2D growth on the flat surfaces, (2) arbitrary combination of two constituent QWs with different thickness, which enables us to study dimensional crossover of the electronic states in the quantum wires, and (3) feasibility of strong 1D confinement in small-size quantum-wire structures.

Meanwhile, the difficulty of the CEO method lies in the MBE growth of high-quality GaAs layers on the (110) surface. The MBE growth of GaAs on the (110) surface requires a low substrate temperature between 470 to 510°C, a V/III beam flux ratio (equivalent-flux pressure ratio) of about 60–100, and a growth rate of around 0.3–0.5 μm/h [27], which are quite different from those required for the well-established MBE growth of GaAs layers on the (001) surface. Also, allowable ranges in growth conditions for growing high-quality layers on the (110) surface are narrower than those in the (001) growth.

2.2   Micro-PL Imaging and Spectroscopy Setup to Characterize T Wires

Figure 3 shows the micro-PL setup used in this study to characterize optical properties of quantum wires and constituent QWs [34]. The micro-PL setup consists of a microscope objective lens, an electrically cooled CCD camera for PL imaging, a monochromator with a liquid-nitrogen-cooled CCD camera for PL spectroscopy, and a monitoring CCD-TV camera with a tungsten illumination lamp. Samples are placed in a liquid-helium continuous flow cryostat so that the overgrowth surface faces to the objective lens as shown in Fig. 4.

In this system, two kinds of photoexcitation are available. One is point photoexcitation. In the point-excitation mode, light from an excitation laser is coaxially focused, through the objective lens, into a near diffraction-limited spot of about 0.8 μm diameter on the sample surface. The other is uniform photoexcitation. In the uniform-excitation mode where a defocusing lens is inserted into the optical path, light from the laser uniformly illuminates the whole of the sample surface.

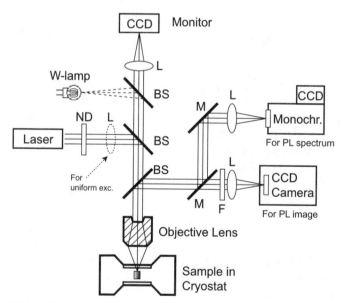

Fig. 3. Schematic of micro-PL setup used in this study. M; mirror, L; lens, BS; beam splitter, ND; neutral density filter, F; sharp-cut filter. Objective lens has numerical aperture of 0.5, nominal magnitude of ×40, and working distance of 10 mm

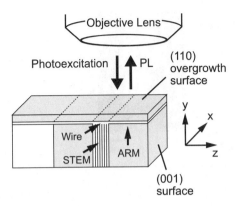

Fig. 4. Configuration of the sample in the micro-PL measurement. The sample is placed in the cryostat so that the overgrowth surface faces the objective lens in the backward scattering geometry

PL from the sample is collected by the same objective lens through an optical glass window of 1.5-mm thickness in the backward scattering geometry as shown in Fig. 4, and is introduced into the CCD camera or the monochromator.

2.3   PL of T Wires Grown by the Original CEO Method

In the T wires grown by the CEO method, novel properties inherent to their one-dimensionality, such as enhanced exciton binding energy and concentrated oscillator strength have been revealed [13]. However, it was recently revealed that the electronic states in the T wires are localized at a low temperature due to monolayer thickness fluctuation and act as a set of QDs [23].

Figure 5 shows a micro-PL spectrum and a micro-PL image measured at 4.8 K of a T-wire sample grown by the original CEO method. The sample has 200 periods of T wires consisting of 200 periods of 5.2-nm thick GaAs stem wells separated by 31-nm thick $Al_{0.3}Ga_{0.7}As$ barrier layers and a 4.8-nm thick GaAs arm well. The total layer thickness of the T-wire region is 7.2 μm. The CEO growth was performed under the typical growth conditions optimized for the (110) GaAs surface shown above [27]. The details of the sample design and preparation are given in [35]. Stability of 1D excitons confined in the T wires is characterized by the lateral confinement energy that is defined by the energy difference between the PL peak position of the wire and those of the constituent QWs. From the PL spectrum in Fig. 5, the lateral confinement energy of the wire is estimated to be 16 meV in this T-wire structure. Larger confinement energies of 35 and 34 meV have been obtained in 5-nm scale GaAs T wires with AlAs barriers and 3.5-nm scale $In_{0.17}Ga_{0.83}As$ T wires with $Al_{0.3}Ga_{0.7}As$ barriers, respectively [4,36].

Note in Fig. 5a that the linewidth of the PL peak from the wire was 7.3 meV. Similar PL linewidths of the T wires have been reported separately

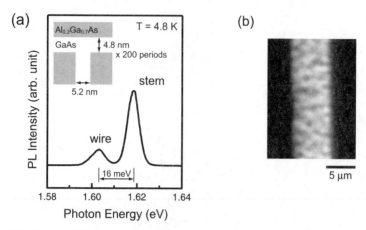

Fig. 5. (a) PL spectrum with point excitation and (b) PL image with uniform excitation, measured at 4.8 K, of equally spaced two-hundred T-wire sample grown by the original CEO method. The inset in (a) shows the structure of each T wire formed in the 200-wire sample

by some research groups [3–5,16,28–31]. Such a broad PL linewidth implies the existence of unavoidable large structural inhomogeneity in the T-wire structure. Figure 5b shows the PL image of the 200 T wires under uniform excitation taken at the PL peak energy of the wire selected by a bandpass filter. Spatially inhomogeneous PL intensity in Fig. 5b also suggests the existence of large structural inhomogeneity.

3   Interface Roughness and Modulated Electronic States in (110) GaAs QWs

To clarify the origins of the structural inhomogeneity existing in the T-wire structures, we characterize, in this section, interface roughness and local electronic states formed in GaAs QWs grown on (110) cleaved surfaces by using high-resolution micro-PL imaging and spectroscopy technique assisted by a solid immersion lens (SIL) [37].

3.1   Preparation of (110) GaAs QWs

5-nm thick (110) GaAs single-QW (SQW) samples with AlAs barriers were grown with MBE on the cleaved (110) edge of a (001) GaAs substrate [37]. The procedure of the sample preparation is as follows. After cleavage in air, the substrate was loaded into the MBE chamber and the top oxide layer was removed at 580°C. Then, 200- (or 500-) nm GaAs and GaAs/AlAs superlattice buffer layers, a 5-nm thick GaAs SQW sandwiched by 10-nm thick AlAs barriers, and a 10-nm thick GaAs cap layer were successively grown without growth interruption at a substrate temperature between 490 to 510°C, a V/III flux ratio of 30, and a GaAs growth rate of 0.35 μm/h.

For comparison, a 5-nm thick GaAs SQW of the same structure with AlAs barriers was grown on a GaAs (001) substrate under the optimal growth conditions with a substrate temperature of 605°C, a V/III flux ratio of 5, a GaAs growth rate of 1.0 μm/h for (001) surface, and a growth interruption time at the GaAs surface of 45 s.

3.2 Macro-PL of the (110) GaAs QWs

The solid and dashed curves in Fig. 6 show the macro-PL spectra of 5-nm thick GaAs/AlAs SQWs grown on the (110) cleaved edge and the (001) substrate, respectively, measured at 77 K under photoexcitation by a He-Ne laser with a spot size of about 100 μm.

The difference in PL peak energy between two GaAs/AlAs SQWs with the same thickness was due to anisotropy of the heavy-hole mass. Note that the PL spectrum of the (110) SQW has a broader linewidth than that of the (001) SQW. The full width at half maximum (FWHM) was 27 meV for the (110) SQW, and 7.2 meV for the (001) SQW. The PL linewidth of 7.2 meV in the (001) SQW was smaller than the energy separation of about 15 meV in the quantization energy due to a one-monolayer (1-ML) difference in well thickness. On the other hand, the PL linewidth of 27 meV in the (110) SQW was greater than the energy separation corresponding to a 3-ML difference

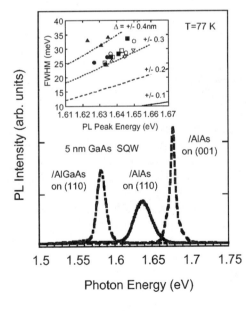

Fig. 6. Macro-PL spectra of a 5-nm thick (110) GaAs/AlAs SQW (*solid curve*), a (001) GaAs/AlAs SQW (*dashed curve*), and a (110) GaAs/Al$_{0.3}$Ga$_{0.7}$As SQW (*dash-dot curve*) at 77 K. The inset shows the FWHM of the PL spectra as a function of PL peak energy for seven (110) GaAs/AlAs SQW samples. Each symbol shows a different sample. Lines in the inset indicate the calculated PL-peak-energy separation between QWs with well-thickness deviation of ±0.1 to 0.4 nm from the centered well thickness

in well thickness, which suggests the existence of larger interface roughness in the (110) SQW.

The inset in Fig. 6 shows the FWHM of the PL spectra as a function of PL peak energy for seven (110) GaAs/AlAs SQW samples grown at different substrate temperatures between 490 and 510°C. The same symbols at several PL peak energies show different excitation positions in the case of the same sample. The larger FWHM of the closed triangles as compared to the others was probably due to lower sample quality with slightly reduced growth temperature, because the sample had higher-density triangle-shaped facet structures on the surface as observed under an optical microscope. Lines in the inset indicate the calculated PL-energy difference due to the well-thickness deviations of ±0.1 to ±0.4 nm from the centered well thickness. Note that most samples follow the calculated lines with the deviation being about ±0.3 to ±0.35 nm. This result indicates that the (110) QWs have almost the same quality at growth substrate temperatures within the range of 490 to 510°C, and that a well-thickness fluctuation of about 3.0 to 3.5 MLs (1 ML = 0.2 nm in the (110) surface) exists in the (110) GaAs QW.

The dash-dot curve in Fig. 6 shows the macro-PL spectrum of an additional reference sample of a 5-nm thick GaAs SQW with $Al_{0.3}Ga_{0.7}As$ barriers grown on a (110) cleaved edge under the same growth conditions as those in the case of the (110) SQWs with AlAs barriers. Its FWHM was 13.5 meV and the corresponding well-thickness fluctuation was 0.74 nm or 3.7 MLs, which is similar to that in the (110) GaAs/AlAs QWs, and indicates that the large well-thickness fluctuation in the (110) GaAs QW is independent of the barrier material.

3.3   Micro-PL Spectroscopy of the (110) GaAs QWs

To investigate local electronic states formed in the (110) GaAs QWs, we performed micro-PL spectroscopy with high-spatial and high-spectral resolution under point excitation. For micro-PL measurements with higher spatial resolution, we used a micro-PL setup combined with a SIL, in which spatial resolution was enhanced to 0.4 μm and a spot size of the focused light in the point excitation was reduced to 0.4 μm through the objective lens and the SIL [38–42]. The spectral resolution was also improved to 0.2 meV by using a 76-cm monochromator with a liquid-nitrogen-cooled CCD camera [37].

To obtain local electronic states at a selected position on the sample exactly, we first obtained PL images under uniform excitation as shown in the upper part of Fig. 7a prior to the spectroscopy, and selected an excitation position for local spectroscopy. Then, we changed the excitation mode to point excitation and obtained PL images as shown in the lower part of Fig. 7a and the PL spectra simultaneously.

Figure 7b shows PL spectra measured at 4.8 K at the position shown in Fig. 7a for various excitation power levels. The intensity of each PL spectrum was calibrated in terms of the excitation power and the acquisition time of the

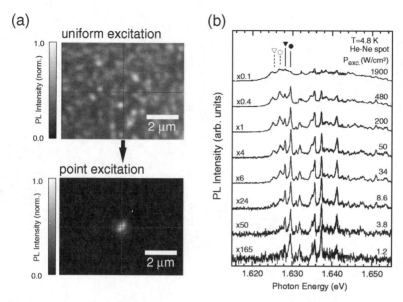

Fig. 7. (a) Micro-PL images observed via SIL under uniform excitation (*upper*) and point excitation (*lower*) in the same region of the (110) GaAs/AlAs SQW. (b) Excitation-power dependence of micro-PL spectra via SIL under point excitation at the position shown in (a). Intensity of each PL spectrum was calibrated on the basis of the excitation power

CCD camera, so that the PL intensity becomes constant if it is proportional to the excitation power.

In the PL spectra under low excitation powers, less than 3.8 W/cm^2, only a few sharp PL peaks were observed. The linewidth of these sharp PL peaks was 0.5 ±0.1 meV.

At higher excitation powers, on the other hand, additional peaks (open symbols) appeared on the low-energy side of the original peaks remaining from low excitation power (solid symbols). To demonstrate this more clearly, we moved the excitation spot to an another position where only a single sharp PL line appeared at the low excitation power. Figure 8a shows the excitation power dependence of the PL spectra at the single PL peak position. As the excitation power was increased, the new peak (open symbol) appeared at the low-energy side of the original peak (solid symbol). Figure 8b shows the intensities of these two PL peaks at various excitation powers. The new emission line increased superlinearly with the excitation power, whereas the original peak showed linear dependence. This result suggests that the sharp emission peak at low excitation powers was due to excitons and the new emission peak was due to biexcitons.

The number of incident photons at the lowest excitation power of 1.2 W/cm^2 in Fig. 7 was 5×10^9 photons/s in the excitation spot area with a diameter of

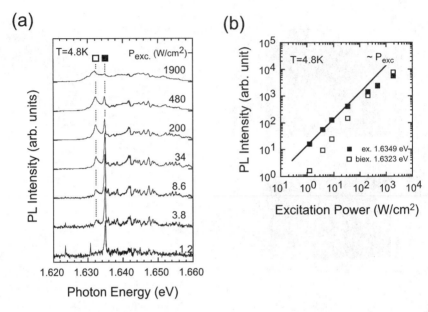

Fig. 8. (a) Excitation power dependence of micro-PL spectra via SIL under point excitation at a position where only a single PL peak is observed, and (b) PL peak intensities of a single excitonic peak at 1.6349 eV and a new emission peak at 1.6323 eV as a function of the excitation power

0.4 μm, which corresponds to an averaged exciton number of 0.04 in the excitation spot assuming that the reflection at the surface is 0.3, the absorption in the SQW is 3%, and the radiative lifetime of excitons is about 0.4 ns. This supports the view that the emission at the lowest excitation level was from single excitons. The new lines begin to appear at an excitation power of about 10 W/cm² corresponding to the averaged carrier number of 0.3. By assuming the Poisson distribution of excitons, we estimated the ratio (n_{xx}/n_x) of probabilities that the spot has the biexciton (n_{xx}) or the exciton (n_x) to be 0.15, which is a reasonable value for the appearance of biexcitons. The broad background PL that appeared at very high excitation power above 1000 W/cm² must originate from interaction between many carriers, or plasma emission.

Note in Fig. 7b that biexciton peaks were observed not only for the lowest-energy PL peak but also for the high-energy peaks, which implies that several local minima in energy existed in the excitation laser spot and that each PL peak observed at the low excitation power was from the excitons in each local minimum in energy. The appearance of such sharp peaks due to reduction of the observation area is quite similar to the results obtained in micro-PL and near-field studies on thin (001) GaAs QWs, where QD states are naturally formed due to the interface roughness of the QWs [43–46]. Therefore, it is concluded that the electronic states are localized due to interface rough-

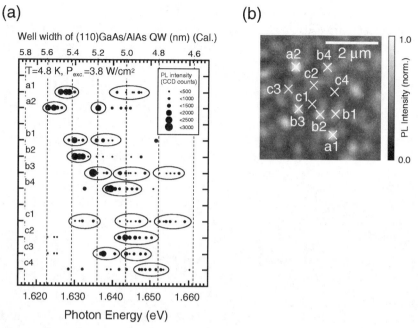

Fig. 9. (a) Peak positions of micro-PL spectra observed at ten different excitation positions on the GaAs/AlAs (110) SQW sample and (b) their excitation position shown as *crosses* on the micro-PL image via SIL under uniform excitation. The size of the dots in (a) represents the PL intensity of each peak. *Vertical dashed lines* in (a) show the calculated PL peak positions for GaAs/AlAs (110) SQWs with well thickness of every ML step

ness and QD-like states are formed in the (110) GaAs QWs. Moreover, the energy separation between the exciton and the biexciton peaks was about 2.5–3.0 meV. This value is similar to the binding energy of biexcitons obtained in the naturally formed QD states in the (001) GaAs QWs [45] and that in the self-organized QDs [47], which also supports our interpretation.

To see the spatial distribution of the local energy minima in the (110) QW, we observed the PL spectra under weak point excitation at several positions on the sample. Figure 9a shows peak positions of the PL spectra observed at ten different positions; two (a1 and a2) at bright PL positions, four (b1–b4) at middle intensity positions, and four (c1–c4) at dark positions, which are marked by crosses in Fig. 9b. In Fig. 9a, the size of the dots represents the PL intensity of each sharp peak. Vertical dashed lines show the calculated PL peak positions for GaAs/AlAs (110) SQWs with different well thickness every 0.2-nm step (1-ML steps).

In Fig. 9a, the bright and dark regions in the PL image of Fig. 9b have different PL peak energy distribution. In brighter regions (a1, a2), relatively strong PL peaks were dominantly observed on the lower-energy side. In darker regions (c1–c4), on the other hand, peaks were located on the higher-energy side with weak intensity. This is explained by the spatial distribution of the quantization energy in the (110) QW due to interface roughness and the carrier migration from high-energy states to low-energy states [37].

Note in the energy distribution of PL peaks in Fig. 9a that, as indicated by circles drawn to guide the eye, three to five neighboring PL peaks made groups, and that those groups had energy spacing almost equal to the separation of the PL peak energy in (110) GaAs QWs with 1-ML difference in well thickness as marked by vertical dashed lines. Such a unique energy distribution was explained by considering the following interface model for the GaAs (110) SQW that the top surface of GaAs/AlAs SQW, where AlAs covered the GaAs surface, has large ML terraces with sub-μm to μm width, and the bottom surface, where GaAs covered the AlAs surface, has shorter-scale roughness due to the low mobility of Al atoms [37].

3.4   Interface Roughness in the (110) GaAs QWs and T Wires Grown by the CEO Method

From the results of macro- and micro-PL spectroscopy, we found that interface roughness as large as 3.0–3.5 MLs exists in the (110) GaAs QWs and that the localized QD states are formed due to this large roughness.

Hasen et al. [23] performed micro-PL measurements of a single T wire grown by the CEO method, and observed discrete sharp emission peaks from the wire electronic states. They concluded that these sharp peaks originated from localized QD states due to the monolayer thickness fluctuation of the two constituent QWs; the first growth (001) QW and the CEO (110) QW. From the micro-PL studies of the (110) GaAs QWs described above it could be concluded that large structural inhomogeneity existing in the interface of the (110) GaAs QWs dominantly contribute to the localization of the electronic states in the T wires.

4   Formation of an Atomically Flat Surface on the (110) GaAs Grown by the CEO Method

The most important issue to realize high-quality T wires is to reduce large surface roughness formed on the (110) GaAs layer by the CEO method. To overcome this difficulty inherent to the CEO growth, we developed a growth-interrupt in situ annealing technique on the epitaxial surface after CEO growth. In this section, we explain the newly developed growth-interrupt annealing technique. To confirm the feasibility of this technique on the CEO growth, we characterize the annealed surface by means of AFM in air. We find that by growth-interrupt annealing at a substrate temperature of 600°C for 10 min, the surface roughness on the (110) GaAs layer is dramatically reduced and atomically flat surface is formed. Moreover, on the basis of characteristic step-edge shapes formed on the annealed surfaces, mechanisms of the flat (110) GaAs surface formation and surface evolution during annealing are discussed [48–52].

4.1 Atomic Arrangements of the (001) and (110) GaAs Surfaces

The epitaxial growth conditions required for GaAs layers on the (110) surface are quite different from those on the (001) surfaces as described in the previous section. This difference is qualitatively explained by the difference of the surface atomic arrangement between the (110) and (001) surfaces. Figure 10 shows perspective views of the atomic arrangements of the (110) and (001) surfaces. In Fig. 10, surface reconstruction is ignored. On the (001) surface, Ga and As layers are alternatively stacked and the surface atoms (As atoms in Fig. 10a) in the topmost layer are two-bonded to the alternate atoms (Ga atoms) in the underlying layer. Therefore, the adatoms attached on the surface can make two bonds with the remaining two dangling bonds of the surface atoms. On the (110) surface, on the other hand, each atomic layer consists of an equal number of Ga and As atoms making zigzag chains. Each atom in the topmost layer has three bonds with three alternate atoms (one in

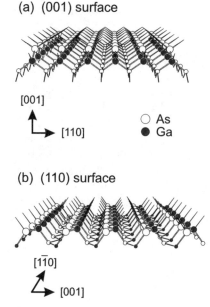

Fig. 10. Atomic arrangements of (**a**) (001) and (**b**) (110) GaAs surfaces. On the (001) surface, the topmost atoms are two-bonded, while on the (110) surface, the topmost atoms have three bonds with three alternate atoms and a single dangling bond

the underlying atomic layer and two neighboring atoms in the same atomic layer) and a single dangling bond that contributes to the epitaxial growth.

In the epitaxial growth of GaAs, the MBE growth is usually performed in a Ga-limited/As-rich growth regime where incident Ga flux determines growth rate of GaAs layer. In this growth regime, the incorporation ratio of Ga atoms is almost unity because desorption (or evaporation) of Ga atoms from the surface can be negligible at a substrate temperature commonly used below at least 650°C [53–55] for both (001) and (110) growth. The crucial point of the epitaxial growth of GaAs in this growth regime is incorporation of As atoms. In the MBE growth of GaAs, As atoms are supplied as As_4 or As_2 molecules on the growth front. It is reported that the incorporation rate of As_4 (and also As_2) molecules is lower on the (110) surface than on the (001) surface at the same growth temperature [56]. If we grow (110) GaAs layers at a substrate temperature as high as 600°C, which is a commonly used substrate temperature in the (001) growth, desorption of As atoms occurs and the growth deviates from the As-rich growth regime. In this condition, unfavorable growths such as a facet formation or formation of Ga droplets occur. Actually, in (110) MBE growth at a substrate temperature as high as 600°C, formation of a large number of facet structures on the epitaxial surface was confirmed [57,58]. Hence, to achieve As-rich condition for growing (110) GaAs layers, one requires a low substrate temperature, high As_4 overpressure, and a low growth rate [59–63]. However, these growth conditions suppress the surface migration of Ga atoms and make the growth surface rough, as shown later.

4.2   Growth-Interrupt in situ Annealing Technique

In the original CEO growth, the overgrowth of the arm well and upper barrier layers of the T wires is continuously carried out on the cleaved edge. Therefore, surface roughness formed on the GaAs arm well layer is embedded in the top heterointerface of the arm well as it is. To reduce the surface roughness formed on the GaAs CEO surface before covering with the barrier layer, we developed a growth-interrupt in situ annealing technique. After the CEO growth on the cleaved edge of the (001) substrates using a growth temperature of 490°C, we interrupt the growth and anneal the sample at an elevated substrate temperature under an As_4 flux irradiation in the MBE chamber [48,49].

4.3   Formation of Atomically Flat CEO Surfaces
by Growth-Interrupt Annealing

After the CEO growth of 5-nm thick GaAs epitaxial layers on the (110) cleaved edge with a growth temperature of 490°C and a growth rate of 0.43 μm/h, we interrupted the growth and annealed the sample under an As_4 flux in the MBE chamber with various annealing times and temperatures.

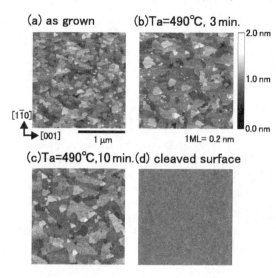

(a) as grown   **(b)Ta=490°C, 3 min.**

2.0 nm

1.0 nm

0.0 nm

[1$\bar{1}$0]

[001]   1 μm   1ML= 0.2 nm

(c)Ta=490°C,10 min.(d) cleaved surface

Fig. 11. AFM images of the as-grown surface of a 5-nm thick (110) GaAs layer grown at 490°C by the CEO method without annealing (**a**), and those of similar layers annealed for 3 (**b**) and 10 min (**c**) keeping the substrate temperature at 490°C. As a reference, the AFM image of a cleaved (110) surface of the substrate with no MBE growth is also shown in (**d**). The observation area is 2 μm × 2 μm and the height scale is 2 nm

It is worth noting that distinct GaAs monolayer-height steps are clearly and reproducibly resolved on the sample surface by AFM in air, as shown below. This indicates that the degradation of the GaAs sample surface due to room-temperature air oxidation is negligible, by which it is suggested that the AFM measurement in air is a powerful tool for investigating the surface morphology of the GaAs epitaxial surfaces on a monolayer scale.

We first characterized the as-grown surface of the 5-nm thick (110) GaAs epitaxial layer grown by the CEO method at 490°C without annealing. This is shown in the AFM image of a 2 μm × 2 μm region in Fig. 11a. As a comparative reference, we show in Fig. 11d an AFM image directly on a bare (110) cleaved surface of the (001) GaAs substrate. The cleave is atomically flat without monolayer steps or islands as expected. The CEO as-grown surface in Fig. 11a is, however, covered with triangular-shaped islands each one or two MLs high and 100 to 200 nm in lateral extent, whose apexes are aligned to the [001] direction. These triangular islands are superposed on underlying monolayer-height terraces forming features generally the same as those previously reported for the conventional MBE growth on (110) substrates [64,65]. From the height analysis of the AFM image, it was found that the mean and peak-to-peak height distribution on the (110) as-grown surface were 1.5 and 5 MLs, respectively.

Note that this as-grown surface of the (110) GaAs CEO layer corresponds to the top interface of the (110) QWs that form T wires studied extensively elsewhere [3–5,16,28–31]. The islands and the relatively large height distribution existing on the as-grown surface observed here nicely accounts for the broad PL linewidth observed in the (110) QWs and T wires grown by the CEO method [3–5,16,28–31,66,67] and the localization of the electronic states in the T-QWRs [23].

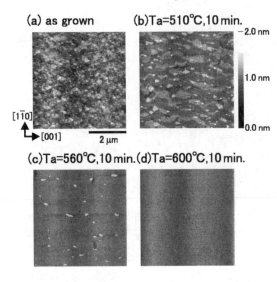

(a) as grown   (b)Ta=510°C,10 min.

−2.0 nm

1.0 nm

[1̄10]

[001]   2 μm   0.0 nm

(c)Ta=560°C,10 min.(d)Ta=600°C,10 min.

Fig. 12. AFM images of the surfaces of a 5-nm thick (110) GaAs layers grown at 490°C by the CEO method and annealed at elevated substrate temperatures for 10 min. The observation area is 5 μm × 5 μm. For comparison, that of the as-grown surface is represented in **(a)** on a 5 μm × 5 μm scale

It is important to clarify whether atom migration tends to flatten the (110) surface or roughen it, for example by step bunching or facet formation. For this purpose, we have characterized the surface morphology of the samples annealed for 3 and 10 min keeping the substrate temperature at 490°C, as shown in Fig. 11b–c, respectively. The observation area was 2 μm × 2 μm.

Note that the surface morphology was improved by annealing. Though the mean and peak-to-peak height distribution were still about 1.6 and 5 MLs, respectively, most islands on the surface were enlarged from sub-μm to μm scale, and the underlying monolayer-height terraces also became larger connecting with one another as the annealing time was increased. Moreover, the 3-min annealed surface shows some smaller islands (< 50 nm in lateral width) that are seen to disappear after the 10-min anneal. The data clearly shows that the migration of atoms (probably Ga) tends to flatten the as-grown (110) surface with time at 490°C [48].

In an attempt to enhance this atom migration, we followed the 490°C MBE growth with anneals at elevated temperatures of 510, 560, and 600°C for 10 min. These are characterized by AFM as shown in Fig. 12b–d, respectively. For this figure, the scanned area shown is extended to 5 μm × 5 μm. For comparison, the AFM image of the 490°C as-grown surface is represented in Fig. 11a also on a 5 μm × 5 μm scale. The gradual undulation of height along the horizontal direction seen in the background of all the images in Fig. 12, particularly in c–d, is an artifact due to the nonlinearlity of the scanning piezo tube in the AFM, and should be ignored.

At all annealing temperatures, particularly as the annealing temperature was increased, the surface morphology was dramatically improved. At the annealing temperature of 510°C, which is 20°C higher than the growth temperature, larger islands of μm-scale were formed and the mean height distribution was reduced to 1.3 MLs, but many sub-μm-scale smaller islands still remained and the peak-to-peak height distribution was still about 4 MLs. The anneals of 560 and 600°C, on the other hand, produced almost flat surfaces. Especially at 600°C, no island step-structures were observed. Indeed an atomically flat monolayer-step-free surface was formed over areas several tens of μm in extent. These results demonstrate that the migration of the surface atoms is enhanced at these higher temperatures and that the (110) GaAs surface is stable even at 600°C for annealing.

We should emphasize that the allowed, and hence optimum, conditions for annealing and growth are different. The growth of (110) GaAs layer requires high As_4-vapor pressure and low substrate temperatures of 470–500°C, because of the low incorporation rate of As atoms to unstable sites on the (110) surface [56]. On the other hand, annealing is not limited by these conditions. The (110) GaAs surface is stable under an As_4 overpressure at substrate temperatures of at least 600°C, where enhanced surface migration of atoms becomes effective in improving the surface morphology of the (110) GaAs epitaxial layer.

It is again stressed that this technique enables us to form high-quality (110) GaAs QWs sandwiched by the two interfaces defined by cleavage and annealing with an atomically flat surface over several tens of μm scale, in which ideal 2D electronic states are expected to be formed.

4.4 Surface Morphology of the Annealed Surface with Fractional Monolayer Coverage

It is now important to characterize the dependence of the surface morphology on the deviation of the amount of GaAs deposition from integer MLs, because fractional ML of GaAs can by no means accomplish an atomically flat surface. For this purpose, a (110) GaAs layer was grown by the CEO method without substrate rotation, but with the CEO surface aligned along the Ga-flux gradient, so that a spatial distribution of GaAs layer thickness by 1%/mm was intentionally introduced [49].

Figure 13 shows AFM images observed at different positions on the (110) GaAs annealed surface. The nominal layer thickness was 6 nm (or 30 MLs), and annealing was again done at 600°C for 10 min. Note in the particular case where exactly 30 molecular MLs of GaAs were deposited over the cleave (denoted as Integer-ML) that an atomically flat surface without any islands nor step edges was formed over an area several tens of μm on a side.

Notice also at other locations along the cleave where the GaAs coverage ended in a fractional monolayer, that the surface morphology is sensitive to the amount of GaAs deposition, and small deviations from the integer-ML deposition cause formation of atomic step-edges and islands or monolayer pits. When the deviation from an integer-ML thickness is +0.1ML or +0.2ML, the GaAs surface is generally atomically flat except for isolated islands shaped like boats (A in Fig. 13), which are 2- or 3-MLs high and elongated along the [001] direction. The existence of such small islands in the interface of the (110)

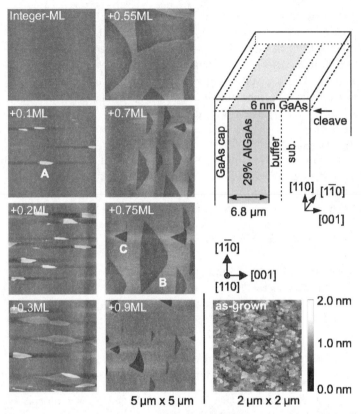

Fig. 13. AFM images of the surface of a 6-nm thick (110) GaAs layer grown on a cleaved (110) edge of a 6.8-μm thick $Al_{0.29}Ga_{0.71}As$ layer on the (001) substrate. The observation area is 5 μm × 5 μm. A schematic sample structure is shown in the inset. The cleaved (110) surface is parallel to the major flat of the (001) substrate wafer. The (110) GaAs layer was overgrown at a substrate temperature of 490°C under As₄ flux. After the overgrowth, in situ annealing of the surface was done at a substrate temperature of 600°C for 10 min. A position with integral-ML deposition is denoted as Integer-ML, and other positions with fractional ML are denoted by deviations of deposition from the integral ML. As a reference, the AFM image of the surface without annealing is labeled as as-grown with an observation area of 2 μm × 2 μm

QWs should cause localized electronic states at low temperatures [43–46]. When the excess coverage becomes +0.3ML, larger islands of 1-ML height are found in addition to boats. At around +0.5ML coverage, the boats disappear as the coalescence of 1-ML-height islands forms connected large terraces. For excess coverage of +0.7ML, a 1-ML-high terrace is extended over the whole surface with a few large isolated 1-ML-deep pits shaped like tropical fish facing toward the [001] direction (B in Fig. 13). At still higher coverage the fish-shaped pits are joined by 2-MLs-deep pits shaped like arrowheads pointing toward the [00$\bar{1}$] direction (C in Fig. 13) [49].

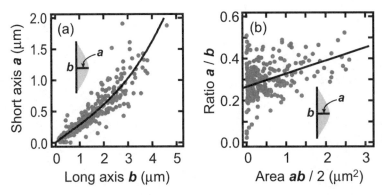

Fig. 14. (a) Distribution of short axis a vs. long axis b in 1-ML deep pits. A fitting curve is drawn to guide the eye. The width and height of a 1-ML deep pit denoted as a and b are measured in the [001] and [1$\bar{1}$0] directions, respectively, as shown in the inset. (b) The a/b ratio vs. the area $ab/2$ of pits. A rising fitting line with increasing pit area shows that large pits have round shapes

4.5 Step-Edge Kinetics on the (110) GaAs Surface during Annealing

Characteristic shapes of the pits and islands formed on the annealed surface at fractional ML regions reveal the atomic-step kinetics during annealing on the (110) surface and illustrate the strong driving force toward flat-surface formation.

Note first that the formed islands and pits are μm-scale in size, and thus are much larger than those observed on *any* (001) GaAs MBE surface [43–46]. This suggests that during the anneal the Ga atom or GaAs molecular mobility is substantially higher on the (110) surface than on the (001) or other surfaces, which is expected from the difference of the surface atomic arrangement between the (110) and (001) surfaces.

To investigate the formation mechanism of such characteristic shapes in more detail, we measured sizes of the fish-shaped 1-ML deep pits, length b along the long axis in the [1$\bar{1}$0] direction and length a along the short axis in the [001] direction. Figure 14a shows length a vs. length b for each fish. A polynomial fitting curves to fourth order shows that a increases superlinearly with b. The data points of a and b are mostly distributed between 0 to 2 μm, and larger fish beyond these sizes are rare. In Fig. 14b, the a/b ratio is plotted as a function of the area $ab/2$, which approximately represents an area of fish by the formula for triangular areas. The data are scattered around a fitting line with a positive slope of 0.03 /μm^2 and an intercept of 0.3. This result indicates that the ratio a/b gradually increases with increasing size of fish. Namely, large fish tend to have round shapes and look fat, while small fish have thinner shapes along the longer b axis [51].

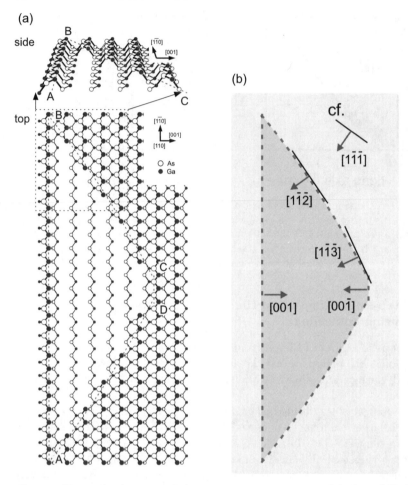

Fig. 15. Top and side views of the atomic arrangement model of a 1-ML-deep pit observed on the annealed (110) surface. The symbol size of those atoms in the topmost atomic layer is enlarged

We model and discuss the kinetics of atomic steps that form flat surfaces and characteristic-shaped islands and pits of Fig. 13 during the growth-interrupt annealing. On the (110) surface at a substrate temperature of 600°C under As_4-vapor overpressure for annealing, desorption and incorporation of As atoms are in equilibrium, while the Ga atoms on the surface have negligible desorption. Thus the surface migration of Ga atoms determines the surface morphology. The fact that islands and pits evolve toward self-similar characteristic shapes on a μm scale indicates that the migration length of the Ga atoms is larger than that scale, and the details of these shapes reflect the relative stability of Ga atoms at the variously oriented step edges on the (110) surface.

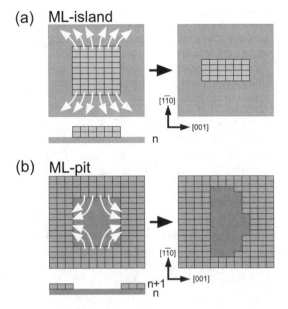

Fig. 16. Atomic-step kinetics of the surface evolution during annealing. Proposed schematic drawings of the evolution for the island and pit structures from an assumed square initial shape are obtained by considering detachment and incorporation of Ga atoms from the ($\bar{1}10$) or ($1\bar{1}0$) edges to the (001) and (00$\bar{1}$) edges at the surface during annealing

In Fig. 15, we reproduced the atomic-step arrangement of the 1-ML deep fish in the AFM image. The longest step edge is the edge A–B consisting of Ga atoms capping As three bonds along the [1$\bar{1}$0] direction. The curved edges come from the edges B–C and A–D formed by Ga atoms capping two As bonds, and the short edge C–D consists of As atoms capping three Ga bonds along the [1$\bar{1}$0] direction. In this discussion we label each step edge on the (110) plane with an index of a plane that is parallel to the edge and perpendicular to the (110) plane. Thus the A–B edge becomes the (001) edge and the C–D edge becomes the (00$\bar{1}$) edge.

The atomic arrangement of Fig. 15 suggests that three-bond sites A–B and C–D are more stable than two-bond edge sites. Furthermore, the edge C–D is less stable than the edge A–B because occasional desorption of an As atom at the C–D step edge leaves two near-neighbor Ga atoms bonded by only two bonds.

On the basis of the difference in the step-edge stability for Ga atoms, we consider the simple surface-evolution model shown in Fig. 16. In the case of 1-ML deep pits, faster detachment of Ga atoms from the less stable ($\bar{1}10$) or ($1\bar{1}0$) edges and incorporation to more-stable sites at (001) and (00$\bar{1}$) edges causes a pit elongated along the [1$\bar{1}$0] direction. This model predicts the time evolution of the pits that the pits should become elongated in the [1$\bar{1}$0] direction and become thinned in the [001] direction as a function of time, because this tends to minimize the number of less-stable two-bond step edges. Note that the shape evolution of larger pits is slower than that in smaller pits, because atom migration in larger pits requires more time. Thus, the dependence of shape on pits area should reflect time evolution. In fact, the result in Fig. 14 shows that smaller fish have thinner shapes, and it is consistent with the above prediction.

For island evolution, on the other hand, the same detachment of the Ga atoms from the less-stable ($\bar{1}10$) or ($1\bar{1}0$) edges causes the remaining island to become elongated along the [001] direction. We here take into account that at the 1-ML high edge, the Ga atom can immediately migrate away from the island after detachment from its initial site, but that in the case of the 2-ML high step edge, a Ga atom in the higher-monolayer edge must first move to an empty site in the lower-monolayer, and only then can it leave the step edge. The kinetics of such a two-step process leads to the conclusion that 2- or higher-MLs islands are more stable than 1-ML islands.

This simple model together with the notion that islands lose Ga atoms to adjacent structures, but pits do not, reproduces the characteristic shapes, relative sizes, and step heights of the islands and pits in the AFM images. This successful modeling of the surface evolution during annealing suggests that the efficient atom migration from unstable two-bond to stable three-bond sites on the (110) surface due to differences in the step-edge stability is the driving force to form an atomically flat surface [50,51].

4.6   First-Principles Calculations of Adatom Migration Barrier Energies on (110) GaAs

The long migration of the adatoms on the (110) surface is also confirmed by first-principles calculations on the migration barrier energy for Ga and As adatoms on a (110) GaAs surface [52]. The calculation for the hopping barrier energy was based on the density functional theory within the generalized gradient approximation. Figures 17a and b show the calculated potential surfaces for the Ga and As adatoms on the GaAs (110) surface, respectively. In Figs. 17c and d, atomic models of Ga and As on the GaAs (110) surface are again shown. Note the difference between the two contour maps a and b. The migration barrier potential surface for Ga adatoms shown in Fig. 17a has low-energy trenches along the [$1\bar{1}0$] direction. Thus, the migration of Ga adatoms is constructed to be 1D along the [$1\bar{1}0$] direction. On the other hand, the migration energy surface of the As adatom in Fig. 17b is 2D.

Table 1 shows the migration barrier energies for Ga and As adatoms on a GaAs (110) surface for the migration to the next stable site in the [$1\bar{1}0$] and [001] directions [52]. For Ga adatoms, the energies are anisotropic: 0.57 eV toward [$1\bar{1}0$] and 0.86 eV toward [001]. On the other hand, the energies for As are 0.57 eV toward [$1\bar{1}0$] and 0.67 eV toward [001], so the difference is only 0.1 eV. These values quantitatively show that the migration of Ga is 1D, while that of As is 2D.

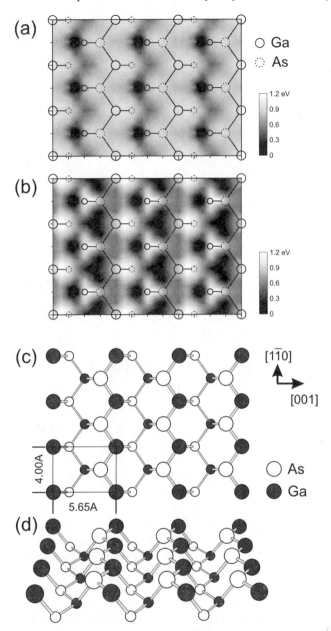

Fig. 17. Contour map of migration barrier energy for (**a**) Ga adatom and (**b**) As adatom on a GaAs (110) surface with the surface atomic configuration. The *dotted* and *solid circles* correspond to As and Ga, respectively. (**c**) and (**d**) Atomic models of a GaAs (110) surface. *Larger* and *smaller circles* correspond to the first and second atomic layers from the surface

Table 1. The migration barrier energy for Ga and As adatoms on the GaAs (110) surface. The calculated barrier energy for Ga on the GaAs (001) $\beta2(2 \times 4)$ surface is also shown for comparison [52]

Atomic species	Migration direction	Barrier energy (eV)
Ga	[1$\bar{1}$0]	0.57
Ga	[001]	0.86
As	[1$\bar{1}$0]	0.57
As	[001]	0.67
Ga	[$\bar{1}$10] on (001)$\beta2$	1.2*
Ga	[110] on (001)$\beta2$	1.5*

* Ref. [68]

For comparison, Table 1 shows the calculated migration barrier energies for the Ga adatom on GaAs (001) $\beta2$ surface [68], which is 1.2 eV toward [$\bar{1}$10] and 1.5 eV toward [110]. It is clear that the barrier energies on GaAs (110) are far smaller than those on GaAs (001). The migration barrier energy of the As adatom on GaAs (110) is also far smaller than that around the Ga-dimer on the GaAs (001) [69], whereas it is similar to the barrier energy value for Ga-rich GaAs (001) $\zeta(4 \times 2)$ surface [70]. Thus, Ga and As adatoms supplied as molecular beams have very small migration barrier energies, and should migrate very easily on the GaAs (110) surface. This explains the experimentally observed formation of a very wide atomically flat (110) surface by growth-interrupt annealing at 600°C for 10 min.

It is interesting to point out, in Fig. 17, that the As adatom is very unstable at the position near the top-layer Ga atom. This means that As is not stable at the site corresponding to the correct site for As in the next atomic layer. The most stable position for an As adatom is the site near As atoms of the topmost surface layer. On the other hand, for a Ga adatom, the stable position is the site near the topmost As. Therefore, for both Ga and As adatoms, the sites near As of the topmost layer are very stable, while those near Ga are unstable. This is most likely because Ga (a group-III element) in the topmost layer has an electronic structure like a closed shell, while As (a group-V element) has excess electrons to make chemical bonds.

The fact that both Ga and As adatoms are stable near As sites and unstable near Ga sites explains the asymmetric shapes like fish for 1-ML deep pits. As shown in Figs. 15 and 16, the lower stability of two-bond step-edges (A–D and B–C) compared with three-bond step-edges (A–B and C–D) causes atom migration from two- to three-bond step-edges and makes the pit elongated along [1$\bar{1}$0]. Note here that both Ga and As adatoms feel repulsion from the Ga-terminated three-bond step-edge (A–B), while they feel attraction from the As-terminated three-bond step-edge (C–D). Thus, adatom density becomes high near the C–D edge. Therefore, for both Ga and As adatoms, the corner sites C and D are the most stable sites for incorporation into the step edge. The repetition of such incorporation processes makes the fish shape of the pit.

4.7 Toward Formation of a Wider Atomically Flat (110) GaAs Surface

From both the experimental results and the theoretical calculations, one can expect that growth-interrupt annealing at even higher substrate temperatures than 600°C is more effective to achieve atomically flat (110) surfaces and most likely to shorten the annealing time. In one experiment, we performed growth-interrupt annealing at a substrate temperature of 650°C on a (110) GaAs surface for 10 min. Figure 18 shows AFM images of the annealed surface of a GaAs layer with nominal thickness of 100 nm or 500 MLs grown on a cleaved edge of a (001) substrate. The GaAs layer was grown without substrate rotation but aligned along the Ga-flux gradient of 1%/mm, which forms 1-ML thickness variation in every 200 μm. If the surface roughness formed during MBE growth is assumed to be proportional to the square root of the layer thickness, the surface roughness of the 100-nm thick layer would be about 22 MLs peak-to-peak. However, on the surface annealed at 650°C, the surface roughness was almost completely removed. The annealed surface showed periodic evolution from an atomically flat to 1-ML deep pits with a period of about 200 μm, but no boat-shaped islands even in the excess region of GaAs deposition were seen. In addition, the fish-shaped pits were more enlarged in size and more elongated along the [1̄10] direction than those observed on the 600°C annealed surface shown in Fig. 13. These features indicate enhancement of the surface migration of Ga adatoms and thus efficient surface flattening at higher temperatures.

Though the present experiment of flat-surface formation is performed on the exact (110) surface prepared by cleavage, one can expect formation of an atomically flat surface over 10 μm also on (110) polished wafers by applying the annealing technique, which directly leads to extensive device application of the (110) surface. Further improvement of this annealing technique is desirable.

5 Fabrication of a High-Quality (110) GaAs QW with Atomically Smooth Interfaces

In this section, we demonstrate fabrication of a 6-nm (110) GaAs QW exactly 30 MLs thick without barrier-well interface roughness using CEO combined with the growth-interrupt annealing. The QW indeed shows a spatially uniform and spectrally sharp PL in micro-PL imaging and spectroscopy [49].

20 μm

Fig. 18. AFM images of a (110) GaAs layer annealed at 650°C for 10 min. The (110) GaAs layer was grown on a (110) cleaved edge of a (001) substrate. The nominal thickness of the GaAs layer was 100 nm or 500 MLs

5.1 Preparation of a (110) GaAs QW with Atomically Smooth Interfaces

The inset in Fig. 19 shows a (110) GaAs QW grown by the CEO method with growth-interrupt annealing. We first formed, by standard MBE on a (001) GaAs wafer, a 300-nm GaAs buffer layer, a 6.8-μm thick $Al_{0.29}Ga_{0.71}As$ barrier layer formed by repeating 20.09 nm $Al_{0.32}Ga_{0.68}As$ layers and 2.55 nm GaAs healing layers with 15 s growth interruption, and a 3000-nm GaAs cap layer. On a cleaved (110) edge of this substrate, a 6-nm thick (110) GaAs layer was grown at a low substrate temperature of 490°C under a high As_4 overpressure by means of the CEO method. To reduce the surface roughness of the (110) GaAs well layer, growth-interrupt annealing was performed on the surface of the GaAs layer at an elevated substrate temperature of 600°C for 10 min under the As_4 molecular flux. To fabricate a QW with atomically flat interfaces, we overgrew an $Al_{0.33}Ga_{0.67}As$ upper barrier layer on the annealed top surface of the GaAs QW layer at a substrate temperature of 490°C. Though the overgrowth thickness was nominally 6 nm on average, a spatial gradation of thickness of 1%/mm was intentionally introduced by aligning the surface along the Ga-flux gradient of MBE without substrate rotation in the same manner as described in the AFM studies on the annealed surfaces in the previous section.

5.2 Micro-PL of the (110) GaAs QW

Figure 19 shows micro-PL images of the GaAs QW at 4.7 K under uniform excitation in the same top-view geometry as shown in Fig. 13. Strong PL comes from the 6.8-μm wide region on $Al_{0.29}Ga_{0.71}As$ where the QW was formed. In the reference QW formed without annealing (denoted as no-annealing), a PL image was spatially inhomogeneous. This is consistent with the previous studies on the related structures [23,37,66,67] and again indicates the existence of a large interface roughness on the top interface of the CEO QW.

On the other hand, in the QW with in situ 600°C annealing, a spatially uniform PL image was observed at the integer-30-ML-thick location (denoted as Integer-ML in Fig. 19), which was also found to extend over several tens of μm in area. This demonstrates the formation of a 6-nm or 30-MLs GaAs QW with no monolayer steps or roughness at *either* AlGaAs interface.

Figure 20 shows the PL spectrum observed at the Integer-ML region under point excitation with a low excitation power of 0.2 nW. For comparison, the PL spectrum for the (110) QW sample without annealing (no-annealing) was observed at the same excitation and detection conditions. We observed a FWHM of 2.1 meV with spectral resolution of 0.9 meV for the annealed QW. This is much narrower than a FWHM of 7.1 meV observed for the reference QW without annealing. In addition, the integrated PL intensity of the annealed QW is almost the same as that of the reference QW as shown in Fig. 20, which means that there is no degradation of PL efficiency in the

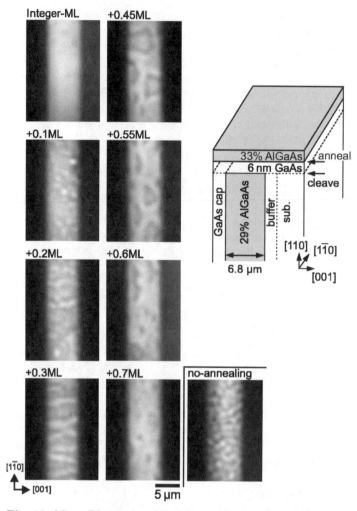

Fig. 19. Micro-PL images of the GaAs QW at 4.7 K under uniform excitation of a He-Ne laser at well thickness positions up to +0.7ML obtained via the (110) surface in the backward scattering geometry. The spatial resolution of the images is 1 μm. For comparison, a PL image of the GaAs QW without annealing is shown as no-annealing

annealed QW. This indicates that during long-term annealing defects such as nonradiative centers or impurities were not introduced into the surface. The origin of the remaining linewidth of 2.1 meV is now not understood. We believe that even our annealed QW still contains additional sources of broadening due, for example, to alloy scattering of the exciton wave function penetrating into the $Al_x Ga_{1-x}As$ barriers where the Al alloy fraction is lo-

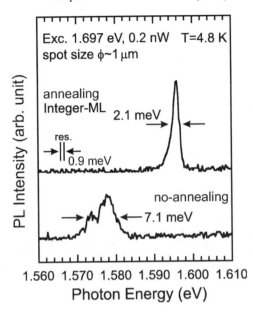

Exc. 1.697 eV, 0.2 nW   T=4.8 K
spot size φ~1 μm

PL Intensity (arb. unit)

annealing
Integer-ML   2.1 meV

res.
→‖←
0.9 meV

no-annealing
→↗↖← 7.1 meV

1.560 1.570 1.580 1.590 1.600 1.610
Photon Energy (eV)

Fig. 20. Micro-PL spectra at 4.8 K of the (110) GaAs QWs grown by the CEO method (**a**) with and (**b**) without growth-interrupt annealing. The excitation power into a spot size of about 1 μm was as low as 0.2 nW. The energy resolution of the micro-PL system used here was 0.9 meV

cally varying and/or interdiffusion of Al and Ga atoms happening at the top heterointerface of the QW during the overgrowth of the $Al_x Ga_{1-x}As$ barrier.

In the other regions of the in situ annealed QW, we see bright PL spots due to boats at 30.2MLs (+0.2ML in Fig. 19), and dark PL profiles shaped like fish at 30.55MLs (+0.55ML in Fig. 19). The AFM patterns of atomic steps observed in Fig. 13 are in fact reproduced in the PL image at each corresponding position of the fractional-MLs deposition. This confirms that the bottom interface formed by the cleavage has no atomic steps, and further that the surface morphology formed on the well layer during annealing and observed by AFM was conserved at the top interface of the QW as the upper barrier material was overgrown.

This conservation of the surface morphology at the QW heterointerface and the resulting formation of the integral-ML-thick QW without interface roughness is also supported by spatially resolved PL spectroscopy associated with the spectrally resolved PL imaging. Figure 21 shows PL spectra at various positions of integer-ML thickness and then as the thickness gradually increases up to an additional 0.7ML. At Integer-ML, only a single PL peak (denoted as n) forms as expected for a QW with integer-ML thickness. As the well thickness is increased by +0.1ML to +0.3ML, emission peaks appear whose energy separation from the peak n corresponds to a +2- or +3-MLs difference in well thickness, but a spectral peak from $n+1$MLs does not. This is consistent with the formation of the boat-shaped islands of 2- or 3-MLs height that we observed in the AFM image. For GaAs depositions in excess of +0.3ML, on the other hand, the peaks $n+2$ or $n+3$ suddenly disappear and a peak corresponding to the $n+1$-MLs thickness appears in the PL spectra.

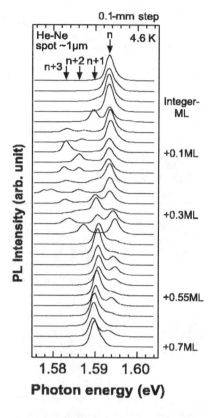

Fig. 21. PL spectra observed at the center of the 6.8-μm wide QW region under point excitation of a He-Ne laser with a spot size of 1 μm. The excitation position was scanned with a step of 100 μm along the [1$\bar{1}$0] direction

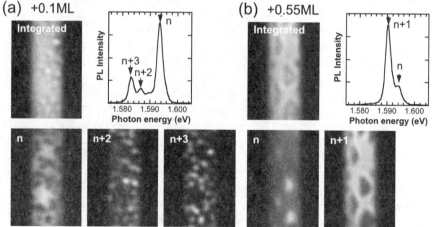

Fig. 22. Integrated and spectrally resolved PL images of the QW and simultaneously obtained PL spectra at 4.7 K under uniform excitation at (**a**) +0.1ML and (**b**) +0.55ML thickness locations. The PL images are resolved at PL peak energies corresponding to the well thickness of n, $n+1$, $n+2$, and $n+3$ MLs

This corresponds to the elimination of boat-shaped islands and formation of a large terrace of 1-ML height.

Figure 22 shows spectrally resolved PL images at the (a) +0.1ML and (b) +0.55ML thickness positions, in which the intensities of the PL peaks denoted as n, $n + 1$, $n + 2$, and $n + 3$MLs are mapped. At the +0.1ML thickness position, the PL image is decomposed into bright spots due to boat-shaped islands at $n + 2$- and $n + 3$-MLs PL peaks and a reverse pattern from the surrounding n-MLs flat region. At the +0.55ML thickness position, the fish-shaped regions with n-MLs thickness are clearly resolved from the surrounding $n + 1$-MLs region.

It is interesting to note that the bright regions of n are smaller in size than the corresponding dark regions of $n + 1$ in Fig. 22b. This directly images the diffusion of QW excitons over the 0.8 μm distance from the higher-energy n-region image edge of the locally narrow QW of fish-shaped pits to the actual step edge of the larger lower-energy $n + 1$ region. Separately, we performed PL imaging under point excitation on the annealed QW and directly evaluated the diffusion length of excitons from the PL images obtained [71]. The diffusion length obtained was as large as 1.3 μm at 4 K, and that gradually increased with the temperature. Such a long-distance path at low temperature is not unreasonable as exciton diffusion is expected to be especially efficient in QWs with atomically smooth interfaces.

The atomically smooth interface formation investigated here by means of the CEO and the in situ annealing method is expected to play an essential role in the realization and future investigation of high-quality quantum wells and wires with ideal 2D and 1D properties.

6   Fabrication of a High-Quality Single-Quantum-Wire Laser Structure and its Lasing Properties

In this section, we fabricate unprecedetedly high-quality single-quantum-wire lasers by the CEO method with growth-interrupt annealing, and demonstrate stimulated emission from the lasers [72,73].

6.1   Preparation of a Single-T-Wire Laser Structure

Using the CEO method combined with the growth-interrupt annealing technique mentioned above, we fabricated a single-quantum-wire laser structure shown in Fig. 23. The T wire consists of a 14-nm thick $Al_{0.07}Ga_{0.93}As$ stem well and a 6-nm thick GaAs arm well, and it was embedded in the core region of a T-shaped optical waveguide structure formed by a 500-nm thick $Al_{0.35}Ga_{0.75}As$ stem layer and a 111-nm thick $Al_{0.10}Ga_{0.90}As$ arm layer [72]. Growth-interrupt annealing was performed for 10 min on the surface of the 6-nm thick GaAs arm well at a substrate temperature of 600°C. Laser bars with an optical cavity length of 500 μm were formed by [1$\bar{1}$0] cleavage from the

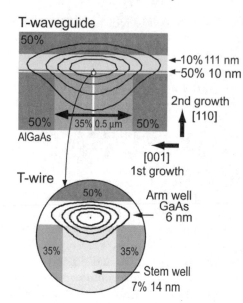

Fig. 23. Schematics of a single-T-wire laser structure fabricated by the CEO method with the growth-interrupt annealing technique. The size of the T wire is 14 nm × 6 nm. The T wire was embedded in a T-shaped optical waveguide

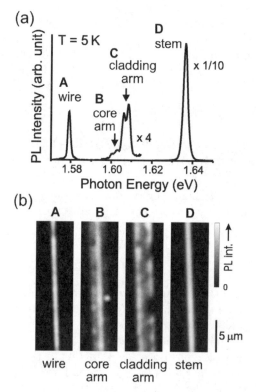

Fig. 24. (a) PL spectrum and (b) PL images of the T-wire laser structure measured under uniform excitation at 5 K. The PL images in (b) were obtained at each PL peak position in (a)

wafer, and the cavity-mirror surfaces were coated with 120-nm and 300-nm thick gold films with an estimated reflectivity of 97%.

6.2   Spatial Uniformity of the Electronic States in the T Wire

Figure 24 shows a micro-PL spectrum and images of the single-T-wire laser structure measured from the top of the overgrowth surface under uniform excitation at 5 K. The PL images were taken at each PL peak selected by a bandpass filter. The photon energy of the excitation light was 1.691 eV and the excitation power was 0.16 mW. By comparing the PL images with the formed T-wire laser structure, we identified peak A as a quantum wire, peaks B and C as arm wells in the core and cladding region of the optical waveguide, respectively, and peak D as a stem well. As shown in Fig. 24b, the PL image of the T wire is spatially uniform over several μm along the wire direction, which indicates that a high-quality single quantum wire was formed.

Figure 25 shows the spatially resolved micro-PL spectra under point excitation scanned over a length of 25 μm of the wire using 0.5 μm steps at 5 K. The lower-energy PL peaks are from the wire, and the higher-energy peaks

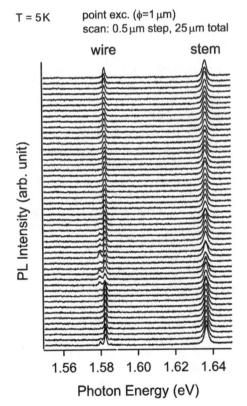

Fig. 25. Spatially resolved PL spectra at 5 K scanned along the wire by steps of 0.5 μm for 25 μm

are from the stem well. No PL from the arm wells was observed because the photogenerated carriers in the arm wells quickly flowed into the T wire.

The main PL peak from the wire observed at 1.582 eV originates from the free excitons, while the small peaks at the lower-energy side of the main peak observed in the middle of the sample are ascribed to the excitons localized in MLs-high islands observed on the annealed surface of the arm well in the fractional GaAs deposition region [48,49]. Both intensity and energy position of the main PL peak of the T wire are spatially uniform over 20 μm. In addition, the PL linewidth of the wire is about 1.3 meV, which is about an order of magnitude narrower than that of the previous T wires fabricated without the annealing procedure. These micro-PL results also confirm that the high-quality T-wire laser was formed.

6.3   Lasing from a Single-Quantum-Wire Laser

Stimulated emission from the single-T-wire laser was measured using optical pumping [72]. Excitation light focused into a filament shape through a set of cylindrical lenses was uniformly incident onto the T-wire laser waveguide from the top of the overgrowth surface, and the light emitted from the edge of the laser cavity was detected.

Figure 26 shows stimulated emission spectra obtained at various excitation powers at 5 K. The photon energy of the excitation light was 1.6455 eV. Multimode lasing from the wire was observed at an excitation power of 8.3 mW, and then it changed to single-mode lasing at 17 mW. At higher excitation powers, the single-mode emission peak showed a slight red shift with mode hopping. The threshold excitation power of the laser emission from the wire was 5 mW, as shown in Fig. 26b. Single-mode emission was observed up to 40 K, but only multimode lasing was observed at 60 K. Lasing from the stem and arm wells was also observed at higher excitation powers of 17 and 260 mW, respectively. It should be noted that the lasing from the wire was single mode, while that from the stem and arm wells was multimode, which reflects the narrower gain spectra of the T wire compared to those of the stem and arm wells. The lower threshold excitation power in the T-wire laser also supports the narrower gain spectra in the T wire.

The single-mode lasing energy of the wire was 1.577 eV, which was 5 meV lower than the peak energy (1.582 eV) of the spontaneous emission from the 1D free excitons of the wire shown in Fig. 25, and there was no overlap between the lasing peak and the 1D free-exciton PL peak. This implies that the lasing gain occurred in the ground state of the T wire but was not due to the recombination of free excitons. In micro-PL spectroscopy, a new peak was observed on a lower-energy side of the free-exciton PL peak at a higher excitation power. This new peak increased in intensity with the broadening of its linewidth but the peak showed no energy shift with an increase in the excitation power, which suggests the formation of an electron–hole plasma with strong Coulomb interactions [74,75]. The lasing energy of the wire was

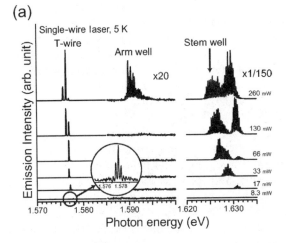

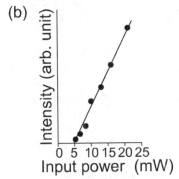

Fig. 26. (a) Stimulated emission spectra of the single-T-wire laser with optical pumping and (b) lasing property of the T-wire laser versus the pumping power

on the lower-energy side of this broad PL emission and the shift in the lasing energy of the wire was less than 2 meV, in contrast to a larger energy shift of 5 meV in the stem well. Therefore, we conclude that the gain for lasing in the quantum-wire laser is due to the strongly Coulomb-correlated electron–hole plasma [74].

7 Concluding Remarks and Future Perspective

By using a modified CEO method combined with the in situ growth-interrupt annealing, we successfully controlled surface morphology of the (110) GaAs layer grown on the cleave and as a result formed an atomically flat surface over several tens of μm in extent in the integral-ML-thick region. We fabricated a (110) GaAs QW with atomically smooth heterointerfaces using this technique and characterized, by means of micro-PL spectroscopy and imaging, optical and structural interface properties in the QW. We confirmed preservation of the atomically smooth (110) surface formed by in situ annealing into the top heterointerface of the QW without any optical and structural degradation. We also fabricated a high-quality single-T-wire laser structure and demonstrated lasing from the T wire. Micro-PL spectroscopy revealed spatial uniformity and high quality of the wire electronic states in the T wire showing sharp PL with a narrow linewidth of 1.3 meV over 20 μm along the wire direction. Lasing from the T wire was achieved with optical pumping at temperatures as high as 60 K, and the origins of gain for lasing in the quantum wires were discussed [74,76].

Successful fabrication of high-quality quantum wires enables us to evaluate novel 1D properties that are hidden behind large structural inhomogeneity in the previous quantum wires. Some of them have already been revealed experimentally. In micro-PL and PL excitation (PLE) spectroscopy of highly uniform T-wire samples, we observed clearly resolved PLs from the ground states and excited states of 1D free excitons and the 1D continuum states [77,78]. It is, for the first time, confirmed experimentally that the singularity of the 1D free-particle density of states in 1D continuum states is suppressed due to strong Coulomb interaction effects, which was predicted theoretically about a decade ago [79,80].

Also, to investigate many-body effects in a 1D electron–hole system in a quantum wire, we observed spectral evolution of PL from the quantum wire with electron–hole pair density using micro-PL measurements. The spectral changes observed indicates that the crossover from a dilute exciton gas to a dense electron–hole plasma occurs gradually via biexcitons and suggests that there are strong biexcitonic correlations in the dense electron–hole plasma in the quantum wire [75]. Many-body effects in a dense 1D electron–hole system is closely related to the gain mechanism for lasing in the quantum-wire lasers. Therefore, further studies both experimentally and theoretically are expected.

Optical response of charged plasma (electron plasma or hole plasma) in doped quantum wires is another interest since understanding of the intriguing difference between charged and electron–hole neutral plasmas in the quantum wires is significantly meaningful from the viewpoint of fundamental physics in 1D systems. We fabricated an n-type modulation-doped single GaAs T wire and observed the evolution of PL spectra with electron density [81]. We observed a clearly resolved 1D charged excitons with large binding energy of 2.3 meV, and large band-gap renormalization and Fermi-edge singularity in a 1D electron plasma, which indicates the crucial importance of Coulomb interaction effects in the doped quantum wire.

From the viewpoint of device application, realization of room-temperature lasing from the T wire is one of the most important targets. In the single-T-wire laser fabricated in this work, lasing was achieved at a temperature as high as 60 K, but not observed at 80 K. In a multiple-T-wire laser structure containing 20 periods of T wires in the active region, lasing at 110 K was achieved, but room-temperature lasing from the T-wire electronic states has not yet been achieved. This is due to the small lateral confinement energy of the T wire to the adjacent QWs. Thus, one direction to realize T-wire devices operating at room temperature is realization of T wires having large confinement energy by optimizing a T-wire structure and/or using adequate materials such as $In_xGa_{1-x}As$ [36,82].

Acknowledgements

The work described here was accomplished in collaboration with Drs L.N. Pfeiffer and K.W. West (Bell Laboratories, Lucent Technologies), Drs J-W. Oh and Y. Hayamizu (University of Tokyo), Professor A. Ishii (Tottori University), and Professors H. Sakaki and T. Someya, and Dr N. Kondo (University of Tokyo). The authors wish to thank all the collaborators.

Part of this research in Japan was supported by the Ministry of Education, Culture, Sports, Science and Technology, Japan and the Japan Science and Technology Agency.

References

1. C. Weisbuch, B. Vinter: *Quantum Semiconductor Structures: Fundamentals and Applications* (Academic Press, San Diego 1991)
2. H. Sakaki, H. Noge: *Nanostructures and Quantum Effects* (Springer-Verlag, Berlin 1994)
3. T. Someya, H. Akiyama, H. Sakaki: Phys. Rev. Lett. **74**, 3664 (1995)
4. T. Someya, H. Akiyama, H. Sakaki: Appl. Phys. Lett. **66**, 3672 (1995)
5. T. Someya, H. Akiyama, H. Sakaki: Phys. Rev. Lett. **76**, 2965 (1996)
6. H. Akiyama, T. Someya, H. Sakaki: Phys. Rev. B **53**, R4229 (1996)
7. H. Akiyama, T. Someya, H. Sakaki: Phys. Rev. B **53**, R10520 (1996)
8. H. Akiyama, T. Someya, H. Sakaki: Phys. Rev. B **53**, R16160 (1996)
9. T. Someya, H. Akiyama, H. Sakaki: Solid State Commun. **108**, 923 (1998)
10. H. Akiyama, S. Koshiba, T. Someya, K. Wada, H. Noge, Y. Nakamura, T. Inoshita, A. Shimizu, H. Sakaki: Phys. Rev. Lett. **72**, 924 (1994)
11. D. Gershoni, J.S. Weiner, S.N.G. Chu, G.A. Baraff, J.M. Vandenberg, L.N. Pfeiffer, K. West, R.A. Logan, T. Tanbun-Ek: Phys. Rev. Lett. **65**, 1631 (1990); **66**, 1375 (1991)
12. D. Gershoni, M. Katz, W. Wegscheider, L.N. Pfeiffer, R.A. Logan, K. West: Phys. Rev. B **50**, 8930 (1994)
13. H. Akiyama: J. Phys.: Condens. Matter **10**, 3095 (1998)
14. R. Ambigapathy, I. Bar-Joseph, D.Y. Oberli, S. Haacke, M.J. Brasil, F. Reinhardt, E. Kapon, B. Deveaud: Phys. Rev. Lett. **78**, 3579 (1997)
15. E. Kapon, D.M. Hwang, R. Bhat: Phys. Rev. Lett. **63**, 430 (1989)
16. W. Wegscheider, L.N. Pfeiffer, M.M. Dignam, A. Pinczuk, K.W. West, S.L. McCall, R. Hull: Phys. Rev. Lett. **71**, 4071 (1993)
17. W. Wegscheider, L. Pfeiffer, K. West, R.E. Leibenguth: Appl. Phys. Lett. **65**, 2510 (1994)
18. J. Rubio, L. Pfeiffer, M.H. Szymanska, A. Pinczuk, S. He, H.U. Baranger, P.B. Littlewood, K.W. West, B.S. Dennis: Solid State Commun. **120**, 423 (2001)

19. S. Watanabe, S. Koshiba, M. Yoshita, H. Sakaki, M. Baba, H. Akiyama: Appl. Phys. Lett. **73**, 511 (1998)
20. L. Sirigu, L. Degiorgi, D.Y. Oberli, A. Rudra, E. Kapon: Physica E **7**, 513 (2000)
21. L. Sirigu, D.Y. Oberli, L. Degiorgi, A. Rudra, E. Kapon: Phys. Rev. B **61**, R10575 (2000)
22. T.G. Kim, X.-L. Wang, R. Kaji, M. Ogura: Physica E **7**, 508 (2000)
23. J. Hasen, L.N. Pfeiffer, A. Pinczuk, S. He, K.W. West, B.S. Dennis: Nature **390**, 54 (1997)
24. F. Lelarge, T. Otterburg, D.Y. Oberli, A. Rudra, E. Kapon: J. Cryst. Growth **221**, 551 (2000)
25. T. Guillet, R. Grousson, V. Voliotis, X.L. Wang, M. Ogura: Phys. Rev. B **68**, 045319 (2003)
26. A. Gustafsson, M.-E. Pistol, L. Montelius, L. Samuelson: J. Appl. Phys. **84**, 1715 (1998)
27. L. Pfeiffer, K.W. West, H.L. Störmer, J.P. Eisenstein, K.W. Baldwin, D. Gershoni, J. Spector: Appl. Phys. Lett. **56**, 1697 (1990)
28. A.R. Goñi, L.N. Pfeiffer, K.W. West, A. Pinczuk, H.U. Baranger, H.L. Störmer: Appl. Phys. Lett. **61**, 1956 (1992)
29. R.D. Grober, T.D. Harris, J.K. Trautman, E. Betzig, W. Wegscheider, L. Pfeiffer, K. West: Appl. Phys. Lett. **64**, 1421 (1994)
30. H. Gislason, W. Langbein, J.M. Hvam: Appl. Phys. Lett. **69**, 3248 (1996)
31. W. Langbein, H. Gislason, J.M. Hvam: Phys. Rev. B **54**, 14595 (1996)
32. R. de Picciotto, H.L. Störmer, A. Yacoby, L.N. Pfeiffer, K.W. Baldwin, K.W. West: Phys. Rev. Lett. **85**, 1730 (2000)
33. W. Wegscheider, G. Schedelbeck, G. Abstreiter, M. Rother, M. Bichler: Phys. Rev. Lett. **79**, 1917 (1997)
34. M. Yoshita, H. Akiyama, T. Someya, H. Sakaki: J. Appl. Phys. **83**, 3777 (1998)
35. T. Someya, H. Akiyama, H. Sakaki: J. Appl. Phys. **79**, 2522 (1996)
36. H. Akiyama, T. Someya, M. Yoshita, T. Sasaki, H. Sakaki: Phys. Rev. B **57**, 3765 (1998)
37. M. Yoshita, N. Kondo, H. Sakaki, M. Baba, H. Akiyama: Phys. Rev. B **63**, 075305 (2001)
38. S.M. Mansfield, G.S. Kino: Appl. Phys. Lett. **57**, 2615 (1990)
39. T. Sasaki, M. Baba, M. Yoshita, H. Akiyama: Jpn. J. Appl. Phys., Part 2 **36**, L962 (1997)
40. M. Yoshita, T. Sasaki, M. Baba, H. Akiyama: Appl. Phys. Lett. **73**, 635 (1998)
41. M. Yoshita, M. Baba, S. Koshiba, H. Sakaki, H. Akiyama: Appl. Phys. Lett. **73**, 2965 (1998)
42. Q. Wu, R.D. Grober, D. Gammon, D.S. Katzer: Phys. Rev. Lett. **83**, 2652 (1999)
43. H.F. Hess, E. Betzig, T.D. Harris, L.N. Pfeiffer, K.W. West: Science **264**, 1740 (1994)
44. A. Zrenner, L.V. Butov, M. Hagn, G. Abstreiter, G. Böhm, G. Weimann: Phys. Rev. Lett. **72**, 3382 (1994)
45. K. Brunner, G. Abstreiter, G. Böhm, G. Tränkle, G. Weimann: Phys. Rev. Lett. **73**, 1138 (1994); Appl. Phys. Lett. **64**, 3320 (1994)
46. D. Gammon, E.S. Snow, B.V. Shanabrook, D.S. Katzer, D. Park: Science **273**, 87 (1996); Phys. Rev. Lett. **76**, 3005 (1996)

47. A. Chavez-Pirson, J. Temmyo, H. Kamada, H. Gotoh, H. Ando: Appl. Phys. Lett. **72**, 3494 (1998)
48. M. Yoshita, H. Akiyama, L.N. Pfeiffer, K.W. West: Jpn. J. Appl. Phys., Part 2 **40**, L252 (2001)
49. M. Yoshita, H. Akiyama, L.N. Pfeiffer, K.W. West: Appl. Phys. Lett. **81**, 49 (2002)
50. M. Yoshita, J-W. Oh, H. Akiyama, L.N. Pfeiffer, K.W. West: J. Cryst. Growth **251**, 62 (2003)
51. J-W. Oh, M. Yoshita, H. Akiyama, L.N. Pfeiffer, K.W. West: Appl. Phys. Lett. **82**, 1709 (2003)
52. A. Ishii, T. Aisaka, J-W. Oh, M. Yoshita, H. Akiyama: Appl. Phys. Lett. **83**, 4187 (2003)
53. R. Fischer, J. Klem, T.J. Drummond, R.E. Thorne, W. Kopp, H. Morkoç, A.Y. Cho: J. Appl. Phys. **54**, 2508 (1983)
54. J.M. Van Hove, P.I. Cohen: Appl. Phys. Lett. **47**, 726 (1985)
55. K.R. Evans, C.E. Stutz, D.K. Lorance, R.L. Jones: J. Vac. Sci. Technol. B **7**, 259 (1989)
56. E.S. Tok, T.S. Jones, J.H. Neave, J. Zhang, B.A. Joyce: Appl. Phys. Lett. **71**, 3278 (1997)
57. W.I. Wang: J. Vac. Sci. Technol. B **1**, 630 (1983)
58. L.T.P. Allen, E.R. Weber, J. Washburn, Y.C. Pao: Appl. Phys. Lett. **51**, 670 (1987)
59. J. Zhou, Y. Huang, Y. Li, W.Y. Jia: J. Cryst. Growth **81**, 221 (1987)
60. A.H. Kean, M.C. Holland, C.R. Stanley: J. Cryst. Growth **127**, 904 (1993)
61. M.C. Holland, C.R. Stanley: J. Vac. Sci. Technol. B **14**, 2305 (1996)
62. D.M. Holmes, J.G. Belk, J.L. Sudijono, J.H. Neave, T.S. Jones, B.A. Joyce: Appl. Phys. Lett. **67**, 2848 (1995)
63. D.M. Holmes, J.G. Belk, J.L. Sudijono, J.H. Neave, T.S. Jones, B.A. Joyce: J. Vac. Sci. Technol. A **14**, 849 (1996)
64. D.M. Holmes, E.S. Tok, J.L. Sudijono, T.S. Jones, B.A. Joyce: J. Cryst. Growth **192**, 33 (1998)
65. M. Wassermeier, H. Yang, E. Tournié, L. Däweritz, K. Ploog: J. Vac. Sci. Technol. B **12**, 2574 (1994)
66. H. Gislason, C.B. Sørensen, J.M. Hvam: Appl. Phys. Lett. **69**, 800 (1996)
67. W. Wegscheider, G. Schedelbeck, R. Neumann, M. Bichler: Physica E **2**, 131 (1998)
68. A. Kley, P. Ruggerone, M. Scheffler: Phys. Rev. Lett. **79**, 5278 (1997)
69. K. Seino, A. Ishii, T. Kawamura: Jpn. J. Appl. Phys., Part 1 **39**, 4285 (2000)
70. K. Seino, W.G. Schmidt, F. Bechstedt, J. Bernholc: Surf. Sci. **507–510**, 406 (2002)
71. J-W. Oh, M. Yoshita, H. Itoh, H. Akiyama, L.N. Pfeiffer, K.W. West: Physica E **21**, 689 (2004)
72. Y. Hayamizu, M. Yoshita, S. Watanabe, H. Akiyama, L.N. Pfeiffer, K.W. West: Appl. Phys. Lett. **81**, 4937 (2002)
73. M. Yoshita, Y. Hayamizu, H. Akiyama, L.N. Pfeiffer, K.W. West: Physica E **21**, 230 (2004)
74. H. Akiyama, L.N. Pfeiffer, M. Yoshita, A. Pinczuk, P.B. Littlewood, K.W. West, M.J. Matthews, J. Wynn: Phys. Rev. B **67**, 041302(R) (2003)
75. M. Yoshita, Y. Hayamizu, H. Akiyama, L.N. Pfeiffer, K.W. West, K. Asano, T. Ogawa: arXiv:cond-mat/0402526 (2004)

82   M. Yoshita and H. Akiyama

76. Y. Takahashi, S. Watanabe, M. Yoshita, H. Itoh, Y. Hayamizu, H. Akiyama, L.N. Pfeiffer, K.W. West: Appl. Phys. Lett. **83**, 4089 (2003)
77. H. Akiyama, M. Yoshita, L.N. Pfeiffer, K.W. West, A. Pinczuk: Appl. Phys. Lett. **82**, 379 (2003)
78. H. Itoh, Y. Hayamizu, M. Yoshita, H. Akiyama, L.N. Pfeiffer, K.W. West, M.H. Szymanska, P.B. Littlewood: Appl. Phys. Lett. **83**, 2043 (2003)
79. T. Ogawa, T. Takagahara: Phys. Rev. B **43**, 14325 (1991)
80. T. Ogawa, T. Takagahara: Phys. Rev. B **44**, 8138 (1991)
81. H. Akiyama, L.N. Pfeiffer, A. Pinczuk, K.W. West, M. Yoshita: Solid State Commun. **122**, 169 (2002)
82. T. Someya, H. Akiyama, H. Sakaki: Jpn. J. Appl. Phys., Part 1 **35**, 2544 (1996)

Recombination Dynamics in $In_xGa_{1-x}N$-Based Nanostructures

Yoichi Kawakami, Akio Kaneta, Kunimichi Omae, Yukio Narukawa, and Takashi Mukai

1 Introduction

Recent progress in epitaxial growth techniques has achieved the successful introduction of indium into gallium nitride, leading to the realization of light-emitting devices based on $In_xGa_{1-x}N/GaN/Al_yGa_{1-y}N$ heterostructures in wurtzite crystal phase, such as dazzling violet, blue, green, and amber light-emitting diodes (LEDs) [1–3], as well as laser diodes (LDs) operated from ultraviolet (350 nm) to blue (480 nm) spectral regions [4–6]. However, further improvement of emission efficiency, reduction in the lasing threshold and extension of the operatable wavelength range is required to expand the application field of such devices. Therefore, it is very important to assess the carrier/exciton recombination processes, through which positive feedback can be made not only to the growth conditions but also to the appropriate design of device structures.

There has been controversy concerning a major role on the modulation of optical transitions in $In_xGa_{1-x}N$-based semiconductors [7]. One important effect, as schematically depicted in Fig. 1a, is exciton localization induced by potential fluctuation because of compositional modulation of In [8–12]. This can be understood as a feature of $In_xGa_{1-x}N$ ternary alloys, where the insoluble tendency of InN is energetically favored in GaN at the growth temperatures [13–16]. Actually, spontaneous formation of In-rich regions on the nanoscopic scale has been reported by several groups [9,17–21]. In this case, large Stokes-like shifts between absorption and emission have been attributed to the exciton/carrier distribution to localized tail states acting as quantum disks or quantum dots depending on the lateral spatial confinement [22–26]. It has been claimed that the high quantum efficiency of $In_xGa_{1-x}N$-based emitters in spite of high threading dislocation density is mostly due to the large localization of excitons/carriers because the pathway to the non-radiative recombination centers is prohibited once they are captured in a small volume [9]. Another important effect is large piezoelectric fields in $In_xGa_{1-x}N/GaN$ quantum wells perpendicular to the (0001)-oriented growth direction as shown in Fig. 1b, which induces the quantum-confined Stark effect (QCSE) [27–31], and the Franz-Keldysh effect (FKE) [32]. As described in the forthcoming section, both effects contribute to the Stokes-like shifts, and to the localization-like behavior, so that conventional macroscopic optical

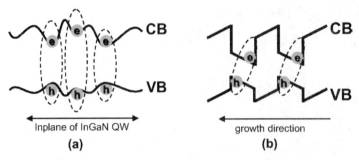

Fig. 1. (a) Potential fluctuation induced by inhomogeneity of indium distribution within $In_xGa_{1-x}N$ QW. (b) Piezoelectric field induced in $In_xGa_{1-x}N/GaN$ QW grown toward (0001) orientation (c-axis)

data do not provide adequate information on the mechanism [33]. Therefore, photoinduced changes of optical absorption spectra were assessed [34–37] to determine which effect plays a major role, by employing white-light pump and probe spectroscopy [38].

A number of reports have recently been appeared on the spatial mapping of luminescence in $In_xGa_{1-x}N/GaN$ quantum wells by optical microscope [39–41], by cathodoluminescence (CL) [42,43], or by scanning near-field optical microscopy (SNOM) [44–53]. It is worth noting that such an SNOM technique is not only useful for the PL mapping, but also applicable to the electroluminescence (EL) mapping under the current-driving condition of LEDs [54,55]. Although the spatial resolution of an optical microscope is typically a few to several hundreds of nanometers that is restricted by diffraction limit, CL has a much smaller diameter exciting electron beam (e-beam). However, the spatial resolution is generally much larger that the e-beam diameter not only because incident electrons spread during the penetration into the sample, but also because generated carriers/excitons diffuse laterally before recombining radiatively or nonradiatively. Another drawback of CL is that both active and cladding layers are photoexcited, while selective photoexcitation to active layers is achieved in optical excitation by tuning incident laser wavelength. The SNOM technique has a potential to attain spatial resolution on the nm scale. Moreover, spatial and temporal dynamics can be detected by combining with time-resolved spectroscopy [50–52].

In this chapter, material parameters are summarized such as bandgap energy and alloy broadening of $In_xGa_{1-x}N$, and piezoelectric field in $In_xGa_{1-x}N/GaN$ quantum wells, and then the general transition models are discussed based on screening of the piezoelectric field, as well as on the localization behavior of excitons/carriers. Finally, detailed results are shown on SNOM-luminescence mapping in an $In_xGa_{1-x}N/GaN$ single-quantum-well (SQW) structure, by which the physical interpretation was made for the recombination mechanism in $In_xGa_{1-x}N$-based nanostructures.

2   Material Parameters of In$_x$Ga$_{1-x}$N

2.1   Bandgap Energies in In$_x$Ga$_{1-x}$N Alloys

In order to assess the optical transitions in In$_x$Ga$_{1-x}$N-based nanostructures, it is essential to estimate the bandgap energies (E_g) of In$_x$Ga$_{1-x}$N ternary alloys, and then take into account the effects of quantum confinement under an electric field, as well as those of localization induced by fluctuation of composition or of the well width. However, there have recently been many papers discussing the exact bandgap energies of In$_x$Ga$_{1-x}$N. This is due to the difficulty in obtaining high-quality InN crystals caused by the low dissociation temperature and high vapor pressure of nitrogen. Although In$_x$Ga$_{1-x}$N-based light emitting devices are in practical use, only a small amount of In up to a few tens of % was successfully incorporated into gallium nitride with sufficient quality in current fabrication technology. Moreover, fabricated In$_x$Ga$_{1-x}$N active layers are highly strained and their layer thicknesses are as small as a few nm. Therefore, the fundamental bandgap of InN was thought to be about 1.89 eV at room temperature (RT) for a long period, where the value had been obtained by the optical absorption measurement on polycrystalline InN grown by a sputtering technique [56]. The bandgap energies of In$_x$Ga$_{1-x}$N have also been estimated in a similar way, and the most commonly cited equation was

$$E_g = 3.42\mathrm{eV}(1 - x) + 1.89\mathrm{eV}x - bx(1 - x) , \qquad (1)$$

where b denotes the bowing parameter, reported values of which scattered in the range between 1.0 eV and 3.8 eV [57–59].

However, PL and absorption measurements of high-quality InN epitaxial layers grown by metalorganic vapor phase epitaxy (MOVPE) and RF-molecular beam epitaxy (RF-MBE) have demonstrated that the fundamental bandgap of InN is about 0.78 eV [60–63], revealing that the appropriate equation should be

$$E_g = 3.42\mathrm{eV}(1 - x) + 0.78\mathrm{eV}x - bx(1 - x) . \qquad (2)$$

In$_x$Ga$_{1-x}$N alloy films with an entire alloy composition have been grown by RF-MBE at about 550°C without noticeable phase separation, and $b = 2.3$ eV [64] was reported by fitting PL and CL peak positions as a function of x-value, while $b = 1.43$ eV [65] was claimed by fitting absorption data, as shown in Fig. 2. Although there still exists, to some extent, the scattering of the value of b, the difference in this case may be contributed from the localization effects, where the luminescence peaks are located on the low-energy side of the absorption edge. Consequently, it has been clarified that In$_x$Ga$_{1-x}$N alloy semiconductors cover not only extend to the ultraviolet and full-visible range but also the infrared reaching to the wavelength used for optical communications [66].

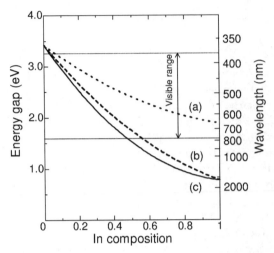

Fig. 2. Bandgap energies of $In_xGa_{1-x}N$ alloys as a function of In content x, calculated assuming various parameters, (**a**) E_g of InN= 1.8 eV, $b = 1.0$ eV [57], (**b**) E_g of InN= 0.78 eV, $b = 2.3$ eV [64], and (**c**) $E_g = 0.77$ eV, $b = 1.43$ eV [65]

2.2 Alloy Broadening Factor in $In_xGa_{1-x}N$ Alloys

The effect of disorder in alloy semiconductors can be classified into two categories, where inhomogeneous broadening is induced by the inevitable nature inherent to alloys, as well as by compositional modulation. The former takes place in a random distribution of alloy composition because of the standard deviation of alloy composition within the exciton Bohr radius. This effect is significant in widegap semiconductors owing to their small exciton Bohr radius. The full width at half maximum (FWHM) of an excitonic transition [$\Delta(x)$] in an $A_{1-x}B_x$ alloy due to this effect is given by the following equation [67]:

$$\Delta(x) = 2\sqrt{2\ln 2}[dE_{ex}(x)/dx]\sqrt{x(1-x)V_0(x)/V_{ex}(x)}, \qquad (3)$$

if a Gussian line shape is assumed, where $E_{ex}(x)$ is the exciton transition energy, $V_0(x)$ is the volume of elementary cell, and $V_{ex}(x)$ is that of an exciton. Zimmermann [68] has derived the relevant exciton volume using a statistical theory expressed by

$$V_{ex}(x) = 8\pi r_B^3(x), \qquad (4)$$

where $r_B(x)$ denotes the Bohr radius of exciton. The $\Delta(x)$ values have been calculated for $In_xGa_{1-x}N$ using r_B values of GaN and InN as 2.9 nm and 7.3 nm, respectively. It should be noted here that no experimental data have been reported for excitonic properties of InN because of the screening of the Coulomb force between electron and hole induced by the high density of

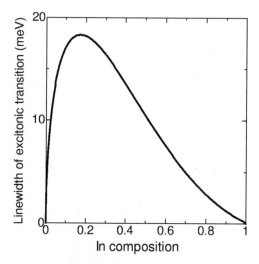

Fig. 3. Theoretical FWHM values of excitonic transition in In$_x$Ga$_{1-x}$N ternary alloys induced by the effect of alloy broadening

residual donors (approximately in the range of 10^{18} cm^{-3} to 10^{20} cm^{-3}) in current InN layers. Therefore, the r_B value of InN has been calculated by

$$r_B = a_0 \epsilon_r \left(\frac{m_0}{m_e} + \frac{m_0}{m_h} \right), \tag{5}$$

where a_0 and ϵ_r are Bohr radius of hydrogen (0.0529 nm) and relative dielectric constant ($\epsilon_r = 15.3$ [69]), and m_0, m_e and m_h are free electron mass, effective mass of electron ($m_e = 0.12 \, m_0$ [70]) and effective mass of hole ($m_h = 1.56 \, m_0$ [71]), respectively. For the x dependence of r_B, a linear variation with x has been assumed.

As shown in Fig. 3, it was found that the FWHM due to this effect is at most 20 meV.

However, observed PL linewidths are much larger than this value, typically from several tens of meV to a few hundred meV depending on samples. This suggests that the compositional modulation, and alloy clustering are induced in samples grown by current growth techniques.

2.3 Piezoelectric Fields in Strained In$_x$Ga$_{1-x}$N Layers

Unlike other semiconductors with zincblende crystal structures grown toward the (001) orientation, a large piezoelectric field is induced in wurtzite-structured In$_x$Ga$_{1-x}$N along the c-axis [(0001)-orientation] [27]. This is not only because of large piezoelectric constants in this direction, but also due to high biaxial-compressive strain in In$_x$Ga$_{1-x}$N layers coherently grown on almost unstrained GaN-based layers. The piezoelectric polarization to the c-axis defined as P_z is given by

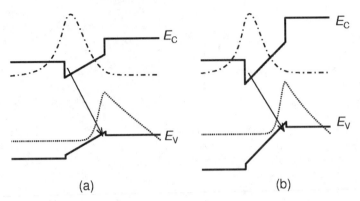

Fig. 4. Band structures as well as wave functions of both electrons and holes in GaN/In$_x$Ga$_{1-x}$N/GaN having 3-nm QW thickness, for **(a)** $x = 25\%$, $E_z = 2.7$ MV/cm, and for **(b)** $x = 50\%$, $E_z = 4.5$ MV/cm

$$P_z = e_{31}\epsilon_{xx} + e_{31}\epsilon_{yy} + e_{33}\epsilon_{zz} , \tag{6}$$

where e_{31} and e_{33} are piezoelectric constants, and ϵ_{xx}, ϵ_{yy} and ϵ_{zz} are strain elements defined by the following equation using a-axis lattice constants of GaN ($a_{GaN} = 0.3189$ nm) and In$_x$Ga$_{1-x}$N ($a_{InGaN} = 0.3189(1 - x) + 0.3548x$ nm), as well as elastic stiffness constants of c_{11} and c_{13}

$$\epsilon_{xx} = \epsilon_{yy} = \frac{a_{GaN} - a_{InGaN}}{a_{GaN}} , \tag{7}$$

$$\epsilon_{zz} = -\frac{2c_{13}}{c_{11}}\epsilon_{xx} . \tag{8}$$

The internal electric field along the c-axis (E_z) is then given by

$$E_z = -\frac{P_z}{\epsilon_r \epsilon_0} , \tag{9}$$

where ϵ_r and ϵ_0 are relative dielectric constants of In$_x$Ga$_{1-x}$N and the permittivity of free space, respectively. Although there still exists wide scattering of reported values of e_{31}, e_{33}, c_{11} and c_{13} of both GaN and InN [72], the E_z value is estimated to be of the order of MV/cm for In$_x$Ga$_{1-x}$N layers with $x = 0.1$–0.2 used for active layers of blue-green emitters [73–75].

In Fig. 4, the band structures of GaN/In$_x$Ga$_{1-x}$N/GaN are depicted assuming that only In$_x$Ga$_{1-x}$N quantum wells are under compressive strain. Due to the internal field, the effective bandgaps as well as oscillator strengths of excitons become much lower than those in the case of flat bands.

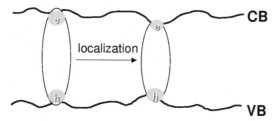

Fig. 5. Potential fluctuation induced by the alloy disorder or by the randomness of the well width

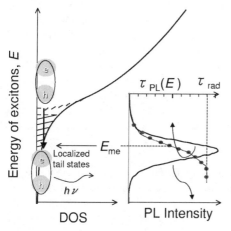

Fig. 6. Density of states (DOS) in localized band structure. PL spectrum, as well as observed PL lifetimes [$\tau_{\mathrm{PL}}(E)$] are plotted schematically

3   General Transition Models

3.1   Localization versus Screening of Piezoelectric Field

Potential fluctuation induces localization of excitons and/or carriers to potential minima in real space. Besides the effect of inevitable alloy broadening, thermodynamical theory shows that the fluctuation of In composition leading to local clustering or phase separation is energetically favored considering the free energy of mixing in In$_x$Ga$_{1-x}$N alloys [13–15]. If In clustering takes place with a size less than several nanometers, localization centers act as quantum dots (QDs) [9,17–21]. Such localization is conspicuous even with a small amount of disorder because of both the small exciton Bohr radius and large bandgap variation with x-value in In$_x$Ga$_{1-x}$N. The model of exciton localization due to potential fluctuation, as well as the density of states (DOS) of the exciton-energy distribution induced by localization are schematically shown in Fig. 5 and Fig. 6, respectively. The observation of large Stokes shift between the absorption edge and the PL peak energy may be interpreted as a result of this effect because radiative recombination takes place at localized

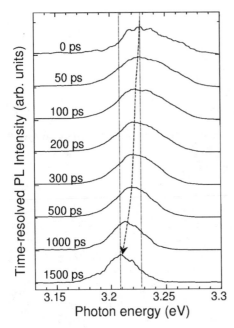

Fig. 7. TRPL spectra of a 0.1-μm thick In$_{0.08}$Ga$_{0.92}$N single epilayer taken at various time delays (0 ps to 1500 ps) after pulsed photoexcitation at 24 K

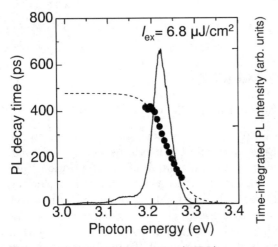

Fig. 8. Time-integrated PL (TIPL) spectrum and the PL decay times [$\tau(E)$] plotted as a function of monitored photon energies taken with a 0.1-μm thick In$_{0.08}$Ga$_{0.92}$N single epilayer at 24 K. The best fitting of $\tau(E)$ was made using values of $\tau_{\rm rad} = 440$ ps, $E_0 = 21$ meV, and $E_{\rm me} = 3.242$ eV

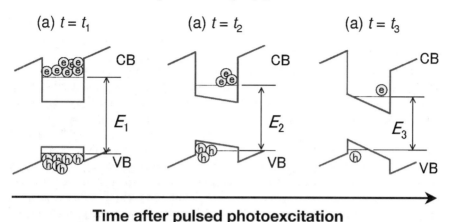

(a) $t = t_1$ (a) $t = t_2$ (a) $t = t_3$

Time after pulsed photoexcitation

Fig. 9. Radiative recombination in a quantum well under a piezoelectric field. Just after pulsed excitation [(a) $t = t_1$], screening of the field is significant due to the large amount of photogenerated carriers. With increasing time, the effect of screening becomes less dominant in accordance with the reduction of carrier density, so that transition energy as well as recombination probability is decreased

tail states, the DOS of which is very small. The recombination dynamics of localized exciton can be assessed by time-resolved PL (TRPL) spectroscopy. If the pulsed photoexcitation is made at continuum energy levels, TRPL peak energies red shift with increasing monitored time after excitation, so that the PL decay time depends on the monitored photon energy with a function of $\tau(E)$. This is because exciton transition is due not only to radiative and/or nonradiative recombination processes but also to the transfer process to the localized levels distributed to the lower-energy side [76]. If the density of localized states is approximated as a single exponential tail with $g_E = \exp(-E/E_0)$ using parameters for the localization depth as E_0, and if the radiative recombination lifetime (τ_{rad}) is constant within the emission band neglecting nonradiative recombination processes, it is possible to estimate $\tau(E)$ by fitting the experimental data with the theoretical equation [76], so that the density of excitons is expressed as functions of energy (E) and time (t) after pulsed excitation with $n(E,t)$

$$\tau(E) = \frac{\tau_{\mathrm{rad}}}{1 + \exp\{(E - E_{\mathrm{me}})/E_0\}} , \tag{10}$$

$$n(E,t) = n_0 \exp\{-t/\tau(E)\} , \tag{11}$$

where E_{me} shows a characteristic energy analogous to mobility edge. Typical examples of TRPL spectra taken with a 0.1-μm thick In$_{0.08}$Ga$_{0.92}$N single epilayer at 24 K are shown in Fig. 7 and Fig. 8, where localization dynamics induced by compositional fluctuation have been revealed [77]. Such characteristics have been observed in wide-bandgap ternary alloys, for example, in CdS$_x$Se$_{1-x}$ [78] and Zn$_x$Cd$_{1-x}$S [79]. However, one has to be careful in

interpreting the recombination mechanism in $In_xGa_{1-x}N$/GaN QW structures by means of TRPL because such localization-like behavior can also be
contributed from the QCSE.

As shown in Fig. 9, electron and hole wave functions in $In_xGa_{1-x}N$ wells
are spatially separated to opposite directions due to the piezoelectric field,
as described in the previous section [27–30]. This leads not only to the reduction of recombination probability of electrons and holes, but also to the
QCSE, where the transition energy is red shifted compared to the case of
flat-band with zero internal electric field. Since the electric field is screened
by carriers either by photoexcitation or by electrical injection, $In_xGa_{1-x}N$-
based LEDs show a blue shift of emission energy with increasing injection
current density. This effect becomes significant for greater well thicknesses
and for higher In mole fractions (x-value). PL lifetime also shows a large well
dependence where it varies from sub-ns to more than a µs if the well width is
increased from 2 nm to 7 nm. Consequently, the PL peak red shifts and the
PL lifetime increases with increasing time after excitaion in accordance with
the reduction of carriers contributing to the screening. It should be noted
that the typical well width of $In_xGa_{1-x}N$-based LEDs and LDs is as small
as 2 to 3 nm in order to make such QCSE as small as possible.

3.2   Photoinduced Change of Optical Density Induced
by Two Major Effects

Two effects, internal electric fields versus potential fluctuation can be separated if the photoinduced change of absorption spectra is characterized by
means of pump and probe spectroscopy, by which the transmission spectrum
of the probe beam detected in the presence of the pump beam $(T + \Delta T)$ is
compared with the spectrum without a pump beam (T) [34–37]. The photoinduced change of the transmission $[\Delta T(\omega, I_{ex}, t_d))]$ depends not only on the
frequency of incident photon (ω) but also on both the photopumping energy
densities (I_{ex}) and time delay between pump and probe beams (t_d) as shown
in Fig. 10. The photoinduced change of optical density $[\Delta OD(\omega, I_{ex}, t_d)]$ is
given by the following equation

$$\Delta OD = \log\left(\frac{T}{T + \Delta T}\right) = \delta\alpha d \times 0.434 , \tag{12}$$

where $\Delta\alpha(\omega, I_{ex}, t_d)$ is the photoinduced change of absorption coefficient,
and d is the total thickness of the absorbing layer. Schematics of the band
structures in $In_xGa_{1-x}N$ QW under FKE and QCSE, and localization effect
are shown in Fig. 11. Since both effects contribute to the broadening of the
absorption edge, it is difficult to separate the two effects by linear absorption
spectra. However, the modification of such optical density due to injected
carriers is different between the two cases. If the time domain of about 1 ps
to ns order is considered, the internal electric field is reduced by the screening

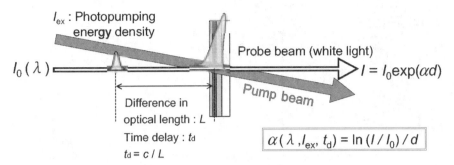

I_{ex} : Photopumping
  energy density

$I_0(\lambda)$

Probe beam (white light)

$I = I_0 \exp(\alpha d)$

Difference in
optical length : L

Pump beam

Time delay : t_d

$t_d = c / L$

$$\alpha(\lambda, I_{ex}, t_d) = \ln(I / I_0) / d$$

Fig. 10. Optical configuration in pump and probe spectroscopy. Time difference between two pulses can be controlled by changing the optical length difference

Table 1. Various types of optical nonlinearities dependent on time domains by classifying positive or negative polarity in ΔOD

Time domain	Origin of nonlinearity	Positive or negative polarity of ΔOD
0–2 ps	Optical Stark effect	ΔOD > 0
	Nonthermal distribution of carriers	ΔOD < 0
	Energy relaxation	ΔOD < 0
1–10 ps	Optical gain	ΔOD < 0
		OD $+ \Delta$OD < 0
1 ps–ns order	Band filling of localized states	ΔOD < 0
	Screening of internal electric fields	ΔOD > 0
ns order–μs order	Screening by trapped carriers	ΔOD > 0
	Thermal effect	ΔOD > 0

effect case (a). Therefore, the optical density (OD) spectra corresponding to absorption spectra (as well as to DOS) become sharp, so the feature of ΔOD is as shown in the figure, where the positive signal is sandwiched between two negative signals. However, in case (b), only the negative signal is observed in ΔOD due to the band filling of localized-tail states. Motivated by this idea, the dynamics of ΔOD was estimated by employing white-light pump-and-probe spectroscopy to In$_x$Ga$_{1-x}$N/GaN multiple-quantum-well (MQW) structures. The effects of (a) the screening and (b) the band filling are the major origins of nonlinear optical density in this time domain, the applicable duration period of which is determined by the carrier recombination times. The optical Stark effect, nonthermal distribution of carriers, energy relaxation and optical gain in the faster time-domain, while the persistent trapped carriers or thermal effects in the slower one can contribute to ΔOD, as shown in Table 1.

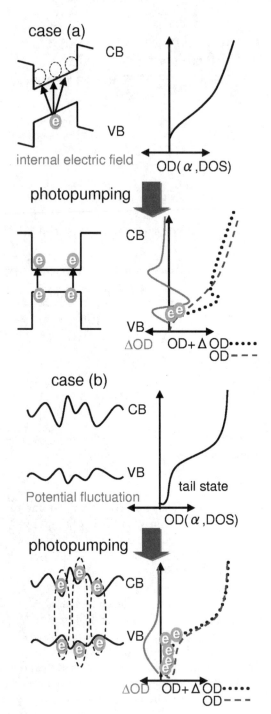

Fig. 11. Schematic band structure and corresponding OD and ΔOD spectra in $In_xGa_{1-x}N$ QW under (**a**) FKE and QCSE effects, and (**b**) localization effect

4   Pump and Probe Spectroscopy on In$_x$Ga$_{1-x}$N Thin Layers and Quantum Wells

The pump and probe spectroscopy depicted in Figs. 12 and 13 was performed for the measurement of temporal behavior of differential absorption spectra. The pump beam with wavelength 370 nm, pulse width 150 fs and repetition rate 1 kHz was formed by passing the output beam from a regenerative amplifier (RGA) to an optical parametric amplifier (OPA). The white light used for the probe beam was generated by focusing part of the output beam from the RGA on a D$_2$O cell. The pulse width of both pump and probe beam was 150 fs. The delay time of the probe beam with respect to the pump beam was tuned by changing the position of the retroreflector that could be controlled by the pulse stage. Since the minimum difference in optical path was 2 µm, a time resolution down to 6.7 fs was achieved. In order to detect the probe beam with spatially uniform carrier distribution in the sample, the focus size of the pump beam (500 µm in diameter) was set to be much larger than that of the probe beam (200 µm in diameter). Furthermore, the probe beam was perpendicularly polarized with respect to the pump beam, and the transmitted probe beam polarized in this direction was detected to avoid the scattered component of the pump beam. Both pump and probe beams were detected by a dual photodiode array in conjunction with a 25-cm monochromator. The four types of samples of In$_x$Ga$_{1-x}$N-based quantum structures used in this study are shown in Fig. 14. They are composed of, respectively, (a) an In$_{0.1}$Ga$_{0.9}$N single layer (30 nm), (b) In$_{0.1}$Ga$_{0.9}$N/GaN (10 nm/10 nm) multiple quantum wells (MQWs) with 3 periods, (c) In$_{0.1}$Ga$_{0.9}$N/GaN (5 nm/10 nm) MQWs with 6 periods, and (d) In$_{0.1}$Ga$_{0.9}$N/GaN (3 nm/10 nm) MQWs with 10 periods. Although the active layers in each sample differs, the total thickness of the active layers in each sample is 30 nm. All the active layers are sandwiched between GaN layers (0.1 µm) and Al$_{0.1}$Ga$_{0.9}$N/GaN (2.5 nm/2.5 nm) superlattices with 100 periods. All layers are grown on GaN buffer layers and sapphire substrates under undoped conditions. Since the PL lifetime of stimulated emission is much shorter than that of the spontaneous one, it was confirmed by means of TRPL that the appearance of stimulated emission at about 3.1 eV under a photoexcitation energy density (I_{ex}) above about 10 µJ/cm^2, corresponding to the threshold carrier density on the order of 10^{18} cm^{-3} for all samples at 10 K.

Figure 15 shows the ΔOD spectra taken at the sample of (a) 30-nm thick In$_{0.1}$Ga$_{0.9}$N single layer at 10 K as a function of time after pumping at 370 nm (3.350 eV) under (1) I_{ex} = 800 µJ/cm^2, and under (2) I_{ex} = 8 µJ/cm^2. The photopumping energy is located below the absorption edge of GaN barrier layers, so that the selective photoexcitation to the active layer was made. The OD spectrum taken under weak photoexcitation, as well as PL spectra under both the same (I_{ex} = 800 µJ/cm^2, stimulated emission) and weak (I_{ex} = 140 nJ/cm^2, spontaneous emission) excitation conditions are shown for reference. In the case of (i) I_{ex} = 800 µJ/cm^2, the carrier density just after the

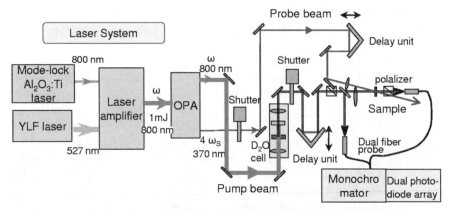

Fig. 12. Optical apparatus used for the pump and probe spectroscopy

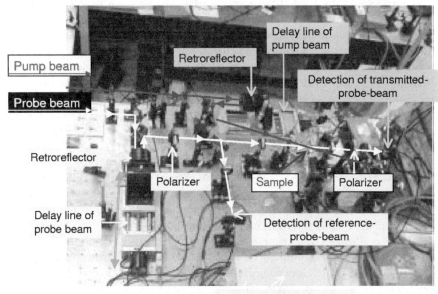

Fig. 13. Picture of beam path and optical component for the pump and probe spectroscopy

excitation is estimated to be about 1×10^{20} cm^{-3}, a negative signal appears at the initiation of time in the whole observed energies. However, a positive signal appears at about 3.2 eV after about 12 ps. These phenomena indicate that the bleaching due to a high density of carriers dominates the spectra initially, but the carrier density is rapidly decreased due to the process of stimulated emission, then photoinduced screening of the internal electric field plays major role on the photoinduced enhancement of OD. In fact, only this

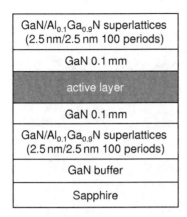

GaN/Al$_{0.1}$Ga$_{0.9}$N superlattices (2.5 nm/2.5 nm 100 periods)
GaN 0.1 mm
active layer
GaN 0.1 mm
GaN/Al$_{0.1}$Ga$_{0.9}$N superlattices (2.5 nm/2.5 nm 100 periods)
GaN buffer
Sapphire

active layer

a. In$_{0.1}$Ga$_{0.9}$N (30 nm)
b. In$_{0.1}$Ga$_{0.9}$N/GaN(10 nm/10 nm × 3)
c. In$_{0.1}$Ga$_{0.9}$N/GaN(5 nm/10 nm × 6)
d. In$_{0.1}$Ga$_{0.9}$N/GaN(3 nm/10 nm × 10)

Fig. 14. Sample structures for pump and probe spectroscopy consisting of four types of samples having different well width, 30 nm, 10 nm, 5 nm and 3 nm. The total thickness of all In$_x$Ga$_{1-x}$N active layers was set to 30 nm in order to make the OD of each sample of comparable value

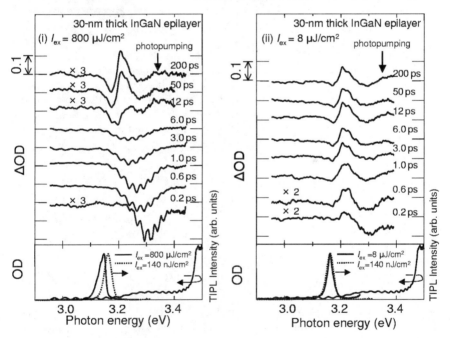

Fig. 15. Variation of ΔOD spectra taken of a 30-nm thick In$_{0.1}$Ga$_{0.9}$N epilayer as a function of time after pumping at 3.350 eV (370 nm) under (i) $I_{ex} = 800$ µJ/cm^2, and under (ii) $I_{ex} = 140$ nJ/cm^2. The OD spectrum under weak photoexcitation, as well as the time-integrated PL (TIPL) spectra under the same and weak excitation conditions are shown at the bottom of each plot

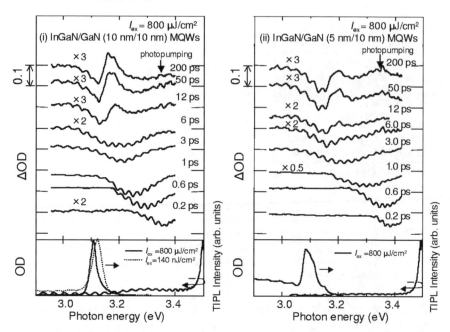

Fig. 16. Variation of ΔOD spectra taken of (**i**) In$_{0.1}$Ga$_{0.9}$N/GaN (10 nm/10 nm) MQWs [sample (b)] and (**ii**) In$_{0.1}$Ga$_{0.9}$N/GaN (5 nm/10 nm) MQWs [sample (c)] as a function of time after pumping at 3.350 eV (370 nm) under I_{ex} = 800 µJ/cm^2. The OD spectra under weak photoexcitation, as well as the TIPL spectra under the same and weak excitation conditions are shown at the bottom of each plot. For the sample (c), the PL spectrum under the weak excitation condition was located at about 2.7 eV that is out of the plotted energy range

positive spectral feature is observed in the whole time range if the pumping energy density is as low as (ii) I_{ex} = 8 µJ/cm^2 where no stimulated emission was observed. It is noted that the same mechanism is also observed in a GaN epilayer, where photoinduced enhancement was observed clearly in excitonic absorption [34].

ΔOD spectra taken at samples of (b) In$_{0.1}$Ga$_{0.9}$N/GaN (10 nm/10 nm) MQWs and of (c) In$_{0.1}$Ga$_{0.9}$N/GaN (5 nm/10 nm) MQWs at 10 K under I_{ex} = 800 µJ/cm^2 are shown in Fig. 16. Similar results with the sample (a) were observed for the sample (b) of 10-nm thick In$_{0.1}$Ga$_{0.9}$N QW. However, the positive feature became less dominant for the sample (c) of 5-nm thick In$_{0.1}$Ga$_{0.9}$N QW.

ΔOD spectra were taken of the sample (d) In$_{0.1}$Ga$_{0.9}$N/GaN (3 nm/1 nm) MQWs (Fig. 17). It was found that carriers photogenerated at 3.350 eV rapidly reach the bottom of the density states within the time scale of several ps. And the important finding is that only a negative signal is observed in the whole spectra, indicating the importance of the band-filling effect compared to the effect of the piezoelectric field.

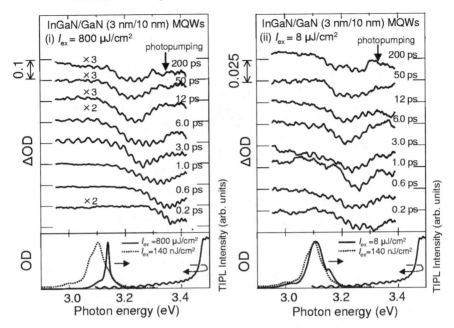

Fig. 17. Variation of ΔOD spectra taken of In$_{0.1}$Ga$_{0.9}$N/GaN (3 nm/10 nm) MQWs [sample (d)] as a function of time after pumping at 3.350 eV (370 nm) under (i) $I_{ex} = 800\ \mu$J/cm^2, and under (ii) $I_{ex} = 140$ nJ/cm^2. The OD spectrum under weak photoexcitation, as well as the TIPL spectra under the same and weak excitation conditions are shown at the bottom of each plot

The following features were clarified, summarizing the results of the pump and probe spectroscopy. In samples (a) and (b), the screening of the internal electric field has a dominant effect on optical transitions after carrier generation. The internal electric field reduces the oscillator strength of the optical transition due to QCSE and FKE. When the photogenerated carriers screen the internal electric field, excitonic absorption is restored. As a result, photoinduced enhancement of the absorption coefficient is observed. The internal electric field strength due to piezoelectric polarization was calculated to be 0.45–1.6 MV/cm using the scattered value of the piezoelectric constants [73,75] as discussed in the previous section. In sample (d), exciton localization has a dominant effect on optical transitions. The density of localized levels is so small that the DOS is readily occupied by the photogenerated carriers. As a result, the broad negative signals of ΔOD are observed. The piezoelectric field dominates over inhomogeneity in samples with layer thickness more than the exciton Bohr radius (3.4 nm) [80,81]. Similar results have been observed at RT though the carrier density to observe the screening of piezoelectric fields in ΔOD is reduced compared to that at low temperature, suggesting that intrinsic carriers activated by thermal energy also contribute persistently to the partial screening of piezoelectric fields.

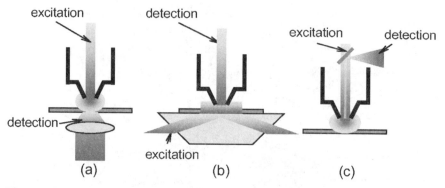

Fig. 18. Three types of optical configuration used for SNOM-PL measurements, (**a**) illumination mode (*I* mode), (**b**) collection mode (*C* mode) and (**c**) illumination–collection mode (*I–C* mode)

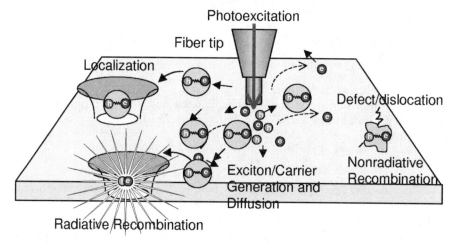

Fig. 19. Schematic of exciton/carrier generation induced by a local photoexcitation by fiber tip, and their recombination dynamics based on diffusion, localization, radiative and nonradiative processes

5   SNOM-Luminescence Mapping Results

5.1   Instrumentation

The SNOM-PL detection has recently been developed as the PL mapping technique, where the optical access in the near-field regime is made through the tip of optical fibers having a very small aperture. Optical configurations in the SNOM-PL technique are schematically illustrated in Fig. 18, where photoexcitation is made in the near field through an optical fiber and the PL signal is detected from the far field with an objective lens in (a) illumination mode (*I* mode), the optical accesses for excitation and detection

(a)
Excitation

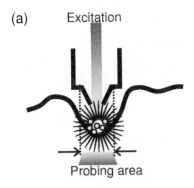

Probing area

(b)
Excitation

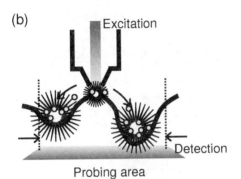

Detection

Probing area

Fig. 20. SNOM-PL detection with (**a**) I–C mode, and (**b**) I mode under potential fluctuation. Spatial resolution is not affected by the carrier/exciton diffusion for (**a**) I–C mode

are opposite to (a) in (b) collection mode (C mode), while both excitation and detection are in the near field in (c) illumination–collection mode (I–C mode). In Fig. 19, the dynamics of carriers/excitons are schematically shown in $In_xGa_{1-x}N$-based layers. Even if they are photogenerated in a very small area by the tip of the optical fiber, some of them diffuse out of the area, and are captured by radiative centers, or by nonradiative recombination centers. The majority of SNOM results for the assessment of $In_xGa_{1-x}N$ have initially been performed in the I mode with an aperture diameter of the optical fiber of 100 nm to a few hundred nm [44–48,51]. However, it is not possible by this mode to determine the true size of localization centers because the spatial resolution is affected by the diffusion process before radiative recombination. This problem can be overcome by means of the I–C mode, where the resolution is solely limited by the diamter of the aperture formed at the fiber tip. The difference between I mode and I–C mode in detecting PL can be understood by taking a look at the probing areas described in Fig. 20.

The SNOM measurements were performed with a NFS-300 near-field spectrometer developed at JASCO Corp. (Figs. 21 and 22) that is capable of photoluminescence (PL) mapping with scanning near-field optical microscopy under I–C mode. Two types of fibers were used in this study; double-tapered Ge-doped-SiO_2 cores with aperture diameter in the range

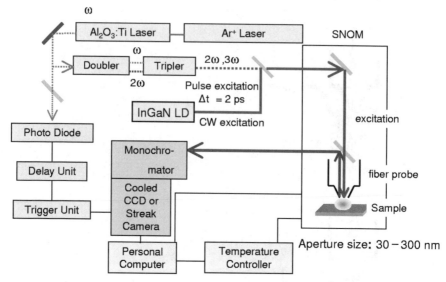

Fig. 21. Schematic of the SNOM system composed of SNOM-head, excitation laser and detection parts, where PL mapping can be performed in both CW and time-resolved modes

from 30 nm to 150 nm, and single-tapered pure-SiO_2 cores with an aperture diameter of 300 nm. A double-tapered structure is fabricated by a multi-step etching technique using hydrofluoric-buffered solution, by which high efficiency of light transmission is achieved compared to that in conventional single-tapered probes. However, one drawback of this structure is the fluorescence of a Ge-doped SiO_2 core that sometimes buries the PL from the sample in the background level if the signal intensity of PL is not strong enough. Therefore, single-tapered pure-SiO_2 fibers fabricated were used for detecting TRPL, where a long exposure time was enabled by the elimination of fluorescence background. Apertures of fiber probes were obtained by applying the mechanical impact on a suitable surface after evaporating Au at the apex. The sample–probe separation was controlled by detecting the amplitude of a dithered probe. The amplitude of this oscillation was less than 1 nm at the first-order resonance frequency of the probe. This amplitude was fed back to control the height of the sample PZT [Pb(Zr,Ti)O_3] stage. As a result, the sample–probe separation was regulated to be 10 nm. The cooling of the samples was performed by flowing cool He gas. The stable measurement was achieved by flowing an appropriated flux of He gas from the bottom to the top of the cryostat. An In_xGa_{1-x}N-based laser diode emitting at 400 nm (developed at Nichia Corp.) was used as the excitation source with the continuous wave (CW) condition. An optical power of 1 mW was coupled to the probe, and about 2 µW was used to illuminate the samples through the probes of both Ge-doped SiO_2 fibers with an aperture diameter

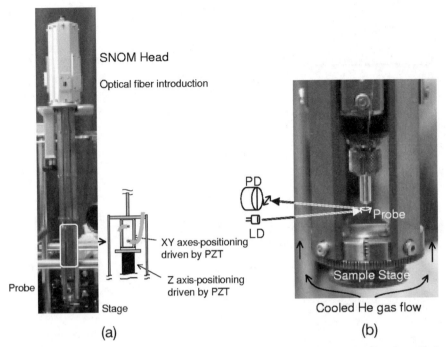

Fig. 22. (a) SNOM-head (whole-view). The position of sample stage is controlled by PZT in xyz directions. (b) The closed view of the sample stage and the probe. The separation between the sample and the fiber tip is controlled by monitoring the friction force using the optical configuration in the figure

of 100 nm and pure-SiO$_2$ cores with an aperture diameter of 300 nm, while about 0.1 μW was used through Ge-doped SiO$_2$ fibers with an aperture diameter of 30 nm. The PL signal was collected by the probe, and was introduced into a monochromator in conjunction with a cooled charge-coupled device (CCD) detector (Roper Scientific, Spec-10:400B/LN). For TRPL measurement, the frequency-doubled mode-locked Al$_2$O$_3$:Ti laser emitting at 400 nm with the pulse width of 2 ps was used as an excitation source. A streak camera (Hamamatsu Photonics, C5680) was used as a detector. It is noted that the selective photoexcitation to the In$_x$Ga$_{1-x}$N active layer was achieved for both measurements. Since the cutoff wavelength of pure-SiO$_2$ fiber is about 1.3 μm, the beam-propagation properties were assessed by measuring pulse width and spectra before and after passing the fiber of 1-m length. It was found that the broadening of the pulse width after transmission is as small as about 10 ps keeping the same wavelength. Therefore, the pure-SiO$_2$ fiber used in this study is suitable for employing a time-resolved SNOM-PL measurement with time-resolution of about 10 ps.

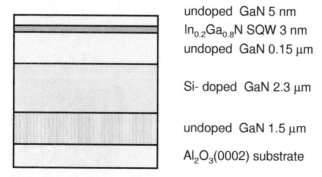

undoped GaN 5 nm
In$_{0.2}$Ga$_{0.8}$N SQW 3 nm
undoped GaN 0.15 μm

Si- doped GaN 2.3 μm

undoped GaN 1.5 μm

Al$_2$O$_3$(0002) substrate

Fig. 23. In$_{0.2}$Ga$_{0.8}$N-SQW structure used for the SNOM measurement

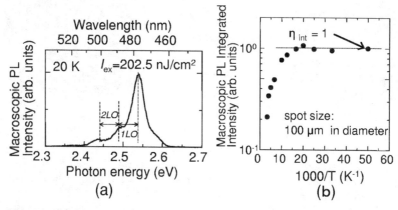

Fig. 24. (a) Macroscopic PL of the sample at 18 K taken with the spot diameter of 100 μm. (b) Macroscopic integrated-PL intensity plotted as a function of inverse of temperature

5.2 SNOM-PL Mapping at Low Temperature under Illumination–Collection Mode

The sample (shown in Fig. 23) is grown on sapphire (0002) substrate by metalorganic chemical vapor deposition (MOCVD), and is composed of a 1.5-μm thick undoped GaN, a 2.3-μm thick n-type GaN:Si, a 3-nm thick In$_x$Ga$_{1-x}$N-SQW active layer ($x \approx 0.2$) and a 5-nm thick undoped GaN layer. The macroscopic PL peak is located at about 480 nm at 18 K as shown in Fig. 24a. LO-phonon side bands are associated on the low-energy side of the main peak. The temperature dependence of the integrated PL-intensity plotted in Fig. 24b shows that the internal quantum efficiency is nearly unity below 50 K because of the suppression of a nonradiative recombination process, and that it is decreased to about 20% at RT.

Figure 25 shows the PL mapping plotted with PL peak intensity as well as with PL peak wavelength at 18 K under a photoexcitation power density

@18K I_{ex}=100 W/cm^2

1.0   2.0   3.0   4.0   5.0   6.0

PL Intensity (arb. units)

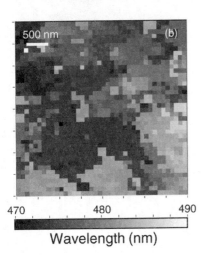

470   480   490

Wavelength (nm)

Fig. 25. SNOM-PL image of an $In_{0.2}Ga_{0.8}$N-SQW structure mapped with (**a**) PL peak intensity and with (**b**) peak wavelength at 18 K under photoexcitation power density of 100 W/cm^2 using a 150-nm aperture fiber-probe

of 100 W/cm^2. The scanning was made in the area of 4×4 µm^2 square with an interval of 100 nm using a 150-nm aperture fiber-probe in I–C mode. It was found that the relative PL intensity fluctuates from 1 to 6, and that the PL peak wavelength is distributed from 470 nm to 490 nm, both of which consist of island-like structures within the range approximately 0.1 to 1 µm.

A clear correlation was observed between PL intensity and wavelength as shown Fig. 26, where the areas of strong PL intensity correspond to those of long PL wavelength (low PL peak energy). The temperature depen-

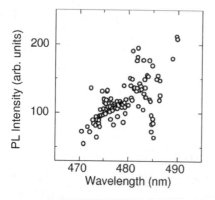

Fig. 26. PL peak intensity plotted as a function of PL peak wavelength from the data of SNOM-PL mapping in Fig. 25

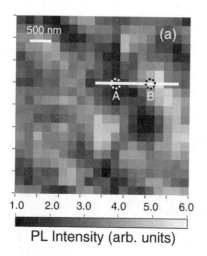

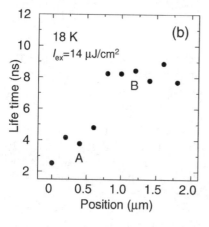

Fig. 27. (a) SNOM-PL peak intensity mapping taken with a pure-SiO₂ fiber with a 300-nm aperture at 18 K. (b) PL lifetimes plotted as a function of the position along the bar shown in (a)

dence of macroscopic PL measurements reveals that the internal quantum efficiency (η_{int}) is nearly unity (0.9–1.0) below 100 K. Moreover, atomic force microscopy (AFM) assessed in situ during the SNOM measurements shows that the root mean square of surface unevenness is as small as 5.1 nm within the scanning area of 4 μm square. The PL peak intensity map (Fig. 1a) shows a relative intensity variation of approximately 1 to 6, corresponding to the η_{int} variation of 0.17 to 1.00 if the maximum is 1.00. If nonradiative recombination alone caused the spatial variation, then the spatially averaged quantum efficiency is estimated to be 0.41 taking into account the area of each PL intensity. This value is much smaller than unity, which is the macroscopic hint as mentioned above. Therefore, nonradiative recombination alone cannot explain the results; diffusion of carrers from the low-intensity to the high-intensity regions must occur.

In order to confirm such a mechanism, TRPL was employed under a SNOM configuration using a pure-SiO$_2$ fiber-probe with a 300-nm aperture. Figure 27a shows the PL image mapped with the PL intensity taken under 100 W/cm^2 with CW condition. TRPL was detected across the white bar drawn in the figure with an interval of 180 nm. The photoexcitation energy density is 14 μJ/cm^2 in this case. PL lifetimes of the emission peak are plotted as a function of position as shown in Fig. 27b. It was found that the short lifetimes (2.5 to 4.8 ns) rapidly jump to the longer ones (7.6 to 9.0 ns) at about 0.75 μm. This position corresponds to the boundary where the PL intensity changes from approximately 2.5 to 5.0. The PL lifetime (τ_{PL}) is expressed by the equation

$$\frac{1}{\tau_{PL}} = \frac{1}{\tau_{rad}} + \frac{1}{\tau_{nonrad}} + \frac{1}{\tau_{trans}} , \tag{13}$$

where τ_{rad} and τ_{nonrad} are radiative and nonradiative lifetimes, respectively, and τ_{trans} represents the transfer lifetime to lower-lying energy levels arising from the localization phenomena. As mentioned above, the term $1/\tau_{nonrad}$ can be neglected at this temperature. Therefore, the shorter lifetimes in the weak PL regions are contributed from the transfer lifetimes. This can be interpreted because PL peak energies of such regions are smaller than other surrounding regions. Figures 28a and b show the TIPL, as well as TRPL spectra as a function of time after pulsed excitation monitored at positions A and B, respectively. The TLPL spectrum in Fig. 28a is composed of two emission bands peaking at 458 nm and 464 nm. The main PL peak at 458 nm decays with a lifetime of 3.8 ns, while the longer peak at 464 nm does so with 6.4 ns. However, for Fig. 28b, the PL band is composed of a single emission peak associated with one LO-phonon replica, and decays with 8.4 ns. Two emission bands with different PL lifetimes in Fig. 28a are probably because the two regions having different energy levels are included within the probing aperture, and the excitons and/or carriers generated at the shorter-wavelength region transfer to the longer-wavelength regions distributed within or out of the aperture. This model is schematically illustrated in Fig. 29.

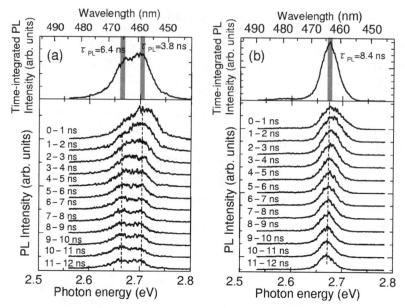

Fig. 28. Time-integrated PL (TIPL) spectra and time-resolved PL (TRPL) spectra as a function of time after pulsed excitation measured at position A (a) and position B (b) at 18 K

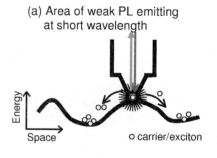

(a) Area of weak PL emitting at short wavelength

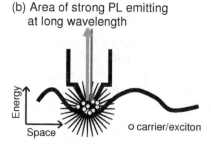

(b) Area of strong PL emitting at long wavelength

Fig. 29. Model of exciton/carrier localization induced by potential fluctuation, and the correlation with the PL intensity monitored by a fiber probe

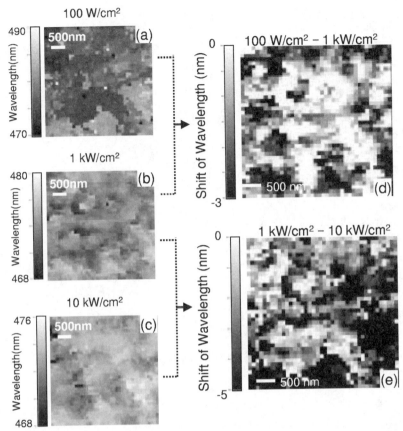

Fig. 30. SNOM-PL mapped with PL peak wavelength under (**a**) $I_{ex} = 100$ W/cm^2, (**b**) 1 kW/cm^2 and (**c**) 10 kW/cm^2. Shift of PL peak wavelength between (**d**) $I_{ex} = 100$ W/cm^2 and 1 kW/cm^2, and between (**e**) $I_{ex} = 1$ kW/cm^2 and 10 kW/cm^2

SNOM-PL spectra were taken at various excitation power density [(a) 100 W/cm^2, (b) 1 kW/cm^2 and (c) 10 kW/cm^2] under CW excitation conditions as shown in Fig. 30. Each monitored position made the shift toward shorter wavelength with increasing excitation power density. However, such a shift is not uniformly distributed as revealed from the mapping of wavelength shifts between $I_{ex} = 100$ W/cm^2 and 1 kW/cm^2 (Fig. 30d), as well as between between $I_{ex} = 1.0$ kW/cm^2 and 10 kW/cm^2 (Fig. 30e). Figure 31 shows the PL peak energies plotted as a function of excitation power for two data points, namely for the weak-intensity region [averaged 100 data point for a value smaller than 25% of PL maximum intensity (I_{max})], and for the strong intensity regions (averaged 100 data point for value larger than 75% of I_{max}). The PL peak energy increases with increasing excitation power in both the strong-intensity region and the weak-intensity region. However, the

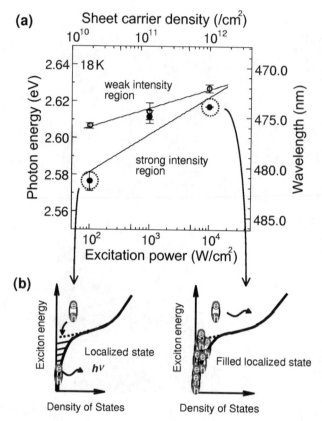

Fig. 31. (a) PL peak wavelength and corresponding peak energy plotted as a function of excitation power density at 18 K, averaged at strong and weak PL-intensity regions. The error bar shows the distribution of the data. (b) The model of band filling in the strong PL-intensity region where the degree of localization is higher than that in the weak PL-intensity region

blue shift is larger in the strong-intensity region than in the weak-intensity region for the same excitation intensity. These results can be explained by assuming that the density of states of localized levels decreases with increasing localization depth. Hence, more filling of the exciton and/or carrier band occurs in the strong-intensity region than in the weak-intensity region for the same excitation intensity. An additional factor that probably contributes to the blue shift in both regions is screening of the piezoelectric field induced by the photogenerated excitons and/or carriers.

In order to assess the spatial distribution of localization centers, CW-PL was performed using a 30-nm aperture probe taken at different positions, as shown in Fig. 32. Several peaks are clearly observed by using a small aperture size fiber-probe and the spectral shape is different. The minimum

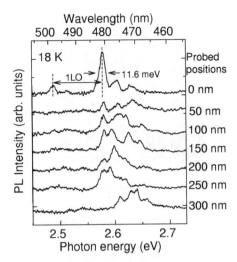

Fig. 32. SNOM-PL spectra taken at each position with a 30-nm aperture under $I_{ex} = 100$ W/cm^2 at 18 K

PL linewidth is about 11.6 meV. This value is one fifth of the macroscopic PL linewidth (about 60 meV), indicating that the macroscopic linewidth is not mainly contributed from the homogeneous broadening due to the interaction with phonons, but from the inhomogeneous one due to potential fluctuations. It is likely that inhomogeneous broadening due to potential fluctuations is still a significant effect on a 30-nm length scale. Therefore, an even smaller PL linewidth might be observed with a smaller aperture size. SNOM-PL intensity mapping was performed with a 30-nm aperture under I–C mode as shown in Fig. 33. The images are taken with four different emission energies, ranging from a low-energy emission component to a high-energy one [(a) 2.560 eV, (b) 2.597 eV, (c) 2.615 eV and (d) 2.636 eV]. The size of the island-like area is in the range from 20 nm to 70 nm for (a) to (c), showing a close distribution. However, islands tend to be connected if the monitored photon energy is the highest (Fig. 33d). It should be noted that such fine structures disappear if monitored under I mode, and that exciton/carrier localization from the high-energy region to the low-energy one was observed by a time-resolved SNOM-PL measurement. Therefore, it is probable that excitons and/or carriers are deeply localized, but each localization center is so closely distributed that they are mobile within the layer as illustrated by the schematic model in Fig. 34. Atomic force microscopy (AFM) assessed in situ during the SNOM measurements shows that the root mean square of surface unevenness is about 3 nm, and that no correlation was found between the unevenness and PL intensity within the scanning area of 250-nm squares. The origin of localization centers thus may be mainly due to the fluctuation of In composition rather than the interface roughness. Cross-sectional TEM

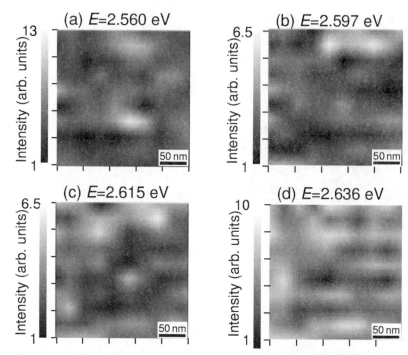

Fig. 33. SNOM-PL intensity images monitored at each emission energy with a 30-nm aperture at 18 K

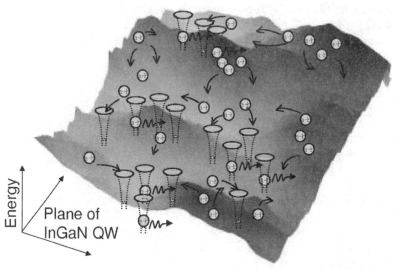

Fig. 34. Recombination model of localized excitons for interpreting SNOM data

observation shows the formation of In-rich QD-like regions about 3 nm in diameter [9]. In compositions in QW and in QDs are estimated to be 20% and 30%, respectively by energy-dispersive X-ray (EDX) microanalysis. The transition energy was thus calculated as a function of the inplane quantum box size assuming that $L_x = L_y$. The result shows that the variation of $L_x (= L_y)$ in the range 2.2 to 3 nm leads to the distribution of localization depth from 70 to 170 meV.

5.3 Multimode SNOM at RT

As described in the previous section, high spatial resolution limited only by the size of aperture is achieved in SNOM-PL mapping under I–C mode. However, a disadvantage of the I–C mode is that it is impossible to detect the signal of radiative recombinations in regions not directly under the probe, while it is possible by a far-field detector of an I-mode configuration. Therefore, it is difficult for the I–C mode to attribute the weak PL intensity to nonradiative recombination processes, or to diffusion of photogenerated carriers outside the detection area of the fiber probe. This is critical for the assessment at RT because the former processes cannot be neglected, unlike at cryogenic temperature.

Focusing on these optical configuration problems, and understanding the importance of collecting different signal simultaneously, we set up a SNOM

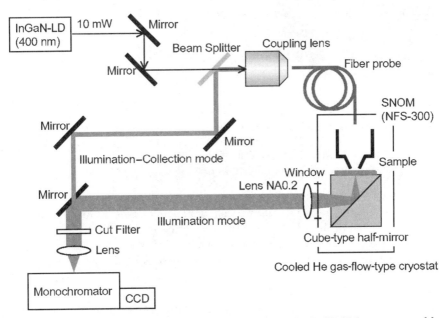

Fig. 35. Schematic of the experimental setup of a multimode SNOM system capable of working simultaneously in I mode and I–C mode to perform CW-PL and TRPL

a *I–C mode* ━ 500 nm

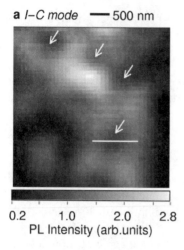

0.2   1.0   2.0   2.8
PL Intensity (arb.units)

b *I mode*

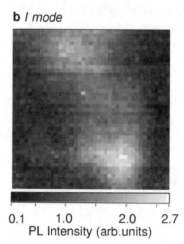

0.1   1.0   2.0   2.7
PL Intensity (arb.units)

Fig. 36. Near-field PL intensity images taken under (**a**) *I–C* mode and (**b**) *I* mode probed with double-tapered fiber (aperture size is 200 nm in diameter) at RT. The excitation power density is 2.5 kW/cm^2 under CW condition. The scanning area is 4 μm × 4 μm with a probing step of 100 nm

apparatus able to operate simultaneously in multiple modes, *I* and *I–C* modes, and designed to probe TRPL spectra in both modes. The multiple measurements taken in this way allowed us to map the PL signal at high resolution and we could clearly discriminate radiative and nonradiative processes in In$_x$Ga$_{1-x}$N-based semiconductors. A schematic experimental setup of this multimode-SNOM is shown in Fig. 35, where optical access in an *I* mode was made from the backside of the sample through a cube-type half-mirror because the sapphire substrate is transparent within the whole spectral range of the detection.

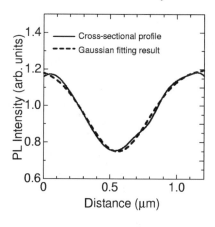

Fig. 37. Cross-sectional profile of PL intensity along the white bar in Fig. 36.(a) The *dashed line* represents the best-fit result using a Gaussian shape

Figure 36 shows the spatial distribution of PL peak intensity under I–C mode and I mode taken in the same scanning area. The excitation power density is 2.5 kW/cm^2 under CW condition. The mean carriers/excitons density for this excitation condition is estimated to be about 5×10^{17} cm^{-3}, considering the absorption coefficient and the light source energy if uniform distribution of carriers is assumed. The scanning area is 4 μm × 4 μm with an interval of 100 nm using a 200-nm aperture fiber-probe. Concerning the I–C mode measurement, it was found that the relative PL intensity fluctuates from 0.2 to 2.8, consisting of island-like structures within the range of approximately 0.3–1 μm. On the other hand, in I-mode measurement, the relative PL intensity fluctuates from 0.1 to 2.7, a value larger than that of the I–C mode. Also at RT, there is no correlation with the PL intensity signal and surface roughness of 3.1 nm within the scanning area of 4 μm^2. It is very interesting to find that differences between two images are observed, where the presence of weak PL-intensity domains (indicated by the arrows) that appear as high PL intensity in I mode. Other regions appear to remain unchanged if observed in the two modes. This behavior can be explained as follows. In the case of domains that appear of weak PL intensity in I–C mode and turn out as high PL intensity in I mode, we believe that the carrier and/or exciton that are photogenerated directly under the optical aperture of the probe, are diffused and localized to out of the I–C-mode probing area, but they remain in the range of the far-field I-mode detector. In the other case, the photogenerated carriers and/or excitons do not migrate further than the I–C-mode probing region, they are presumably captured at nonradiative recombination centers, the origin of which is related to microscopic dislocations and/or to nanoscopic point defects. The cross-sectional profile of PL intensity along the white line in Fig. 36a is plotted in Fig. 37. The FWHM of a Gaussian fitting result of this profile is 550 nm, therefore, the diffusion length to the radiative recombination center is at least 275 nm in this area. It is noted that a similar value is reported by Cherns et al. [82] as the diffusion length in In$_x$Ga$_{1-x}$N-based

(a) *I–C mode* ━━ 500 nm

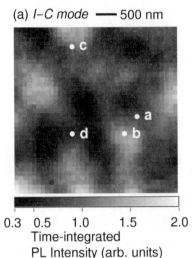

0.3 0.5   1.0   1.5   2.0
Time-integrated
PL Intensity (arb. units)

(b) *I mode*

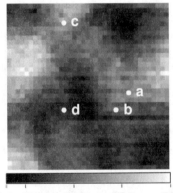

3.4   5.5   11.0   16.5   22.7
Internal quantum efficiency (%)

Fig. 38. Time-integrated near-field PL intensity images taken under *I–C* mode (a) and (b) *I* mode probed with single tapered fiber (pure SiO$_2$, aperture size is 200 nm in diameter) at RT. The excitation energy density is 5.5 µJ/cm^2 under pulsed condition. The scanning area is 3.7 µm×3.7 µm with probing step of 100 nm

quantum structures using the CL spectroscopy technique though this method is used for the characterization of the diffusion to nonradiative recombination centers originating from threading dislocations.

Figure 38a and b shows the PL intensity mapped in *I–C* mode and *I* mode, respectively. TRPL was detected at 4 different positions that are indicated with the letter (a)–(d). These positions were selected as representative

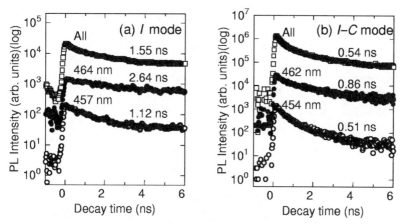

Fig. 39. Decay spectra under (**a**) I mode, and (**b**) I–C mode monitored at spectrally integrated PL intensity (all) and at each PL-energy components. Decay curves are fitted with double-exponential curves, but the fast components of lifetimes are shown in the figure

of 4 different behaviors, (a); relatively weak PL intensity in the I–C mode while stronger PL intensity in I mode, (b); opposite situation to the case of (a), (c); relatively strong PL intensity in both modes and (d); relatively weak PL intensity in both modes. The photoexcitation energy density is 5.5 μJ/cm^2 corresponding to an estimated carrier/exciton density just after photoexcitation to be about 1×10^{18} cm^{-3}. It was found that the PL lifetimes (τ_{PL}) under the I–C mode are always shorter than those in I mode. The difference is significant for the data at (a), where τ_{PL} values of spectrally integrated PL intensity are 0.541 ns and 1.553 ns, for the I–C mode and I mode, respectively, as shown in Fig. 39, both of which are fast components in double-exponential fitting. This assumption is valid because the faster components are dominant comparing to the slower ones, and is necessary to make a simple discussion as described below. The PL lifetime in the I–C mode (τ_{PL-I-C}) is

$$\frac{1}{\tau_{PL-I-C}} = \frac{1}{\tau_{rad}} + \frac{1}{\tau_{nonrad}} + \frac{1}{\tau_{tr-out}} , \tag{14}$$

where τ_{rad} and τ_{nonrad} are radiative and nonradiative lifetimes, respectively, and τ_{tr-out} represents the lifetime of carrier transfer from the area directly under the tip aperture (within the I–C mode probing range, I-mode and I–C-mode detection are both possible) to the external region (in this case only I-mode detection is possible). Since the PL signal is detected in the far-field configuration under I mode, the term $1/\tau_{tr-out}$ can be neglected; the PL lifetime under I mode (τ_{PL-I}) is expressed by

$$\frac{1}{\tau_{PL-I}} = \frac{1}{\tau_{rad}} + \frac{1}{\tau_{nonrad}} . \tag{15}$$

Table 2. Internal quantum efficiency and PL-lifetime in each mode monitored at 4 different positions

Position	$I_{\text{I}-\text{Cmode}}$	$I_{\text{I}-\text{mode}}$	η_{int}	$\tau_{\text{PL}-\text{I}-\text{C}}$ (ns)	$\tau_{\text{PL}-\text{I}}$ (ns)
a	Weak	Strong	21.1	0.541	1.55
b	Strong	Weak	15.0	0.673	0.67
c	Strong	Strong	22.6	0.564	0.568
d	Weak	Weak	9.0	0.552	0.677

It should be noted that this treatment is based on the first-order assumption, where radiative lifetimes as well as nonradiative recombination times in the I mode are averaged to be same as those in the I–C mode. More detailed analysis taking into account the difference in radiative/nonradiative lifetimes is in progress. The TIPL intensity mapped under I mode (Fig. 38b) represents the spatial distribution of internal quantum efficiency. According to the temperature dependence of macroscopic PL measurements it was found that the internal quantum efficiency of this sample is nearly unity (more than 90%) at temperatures less than 50 K. Consequently, the distribution of PL intensities from 0.1 to 2.7 at RT corresponds to η_{int} values ranging from 3.4% to 22.7%. Since the η_{int} value is expressed by

$$\eta_{\text{int}} = \frac{\tau_{\text{nonrad}}}{\tau_{\text{rad}} + \tau_{\text{nonrad}}}, \qquad (16)$$

all recombination lifetimes can be calculated using the experimental data as shown in Table 3. It is evident that the shorter lifetime of $\tau_{\text{PL}-\text{I}}$ probed at position (a) is due to a small $\tau_{\text{tr}-\text{out}} = 0.83$ ns term. This transfer process is probably caused by exciton/carrier localization centers that are local potential minima distributed in the proximity of the tip, but outside of the I-C-mode probing range. This idea is confirmed by examining the time integration of the PL peaks, in the case of position (a) the time-integrated peak is located at 464.2 nm (2.670 eV) under I–C mode, while it is at 461.9 nm (2.683 eV) under I mode, as shown in Fig. 40.

In the point indicated with (b), relatively weak PL intensity in I mode is caused by a transfer process to nonradiative recombination centers distributed in the region external to the I–C probing, as is indicated by the small $\tau_{\text{nonrad}} = 0.79$ ns. Concerning the point in (c), the strong PL intensities in both modes are due to radiative recombinations that mainly take place within the aperture, as shown by a large value of $\tau_{\text{tr}-\text{out}} = 87.1$ ns. Moreover, in the position (d), a weak PL intensity in both modes is due to the large density of nonradiative recombination centers distributed within and outside of the aperture range.

Recombination processes probed at (a)–(d) within inplane potential fluctuations are schematically illustrated in Figs. 41a–d. Based on the dynamics

Table 3. Recombination lifetimes at each position obtained by the calculation of experimental data

Position	$\tau_{\text{tr-out}}$ (ns)	τ_{rad} (ns)	$\tau_{\text{non-rad}}$ (ns)
a	0.83	7.35	1.97
b	4.18	4.49	0.79
c	87.1	2.50	0.73
d	3.00	7.44	0.74

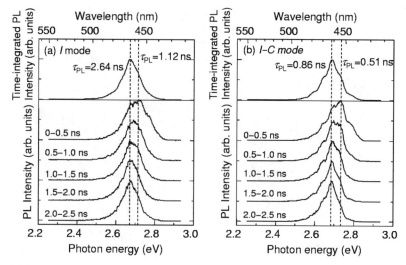

Fig. 40. Time-integrated PL (TIPL) spectra as well as time-resolved PL (TRPL) spectra monitored under (**a**) *I* mode and under (**b**) *I–C* mode. Lifetimes in the figure are experimental decay times monitored at each emission energy shown by *dotted lines*

described above, the transfer, radiative and nonradiative processes taking place are represented in the scheme of Fig. 42. Radiative and nonradiative recombination centers are present all over the sample. However, their densities are inhomogeneous. Higher density of radiative recombination domains act as attractive centers for photogenerated excitons/carriers. The potential energy was estimated by PL peak mapping that we performed separately; in Fig. 42 the dotted lines represent regions where the potential energy is higher. These high-energy lines form a potential ridge that presumably would suppress the carrier/exciton diffusion, creating the carrier dynamics we observed.

120　Y. Kawakami et al.

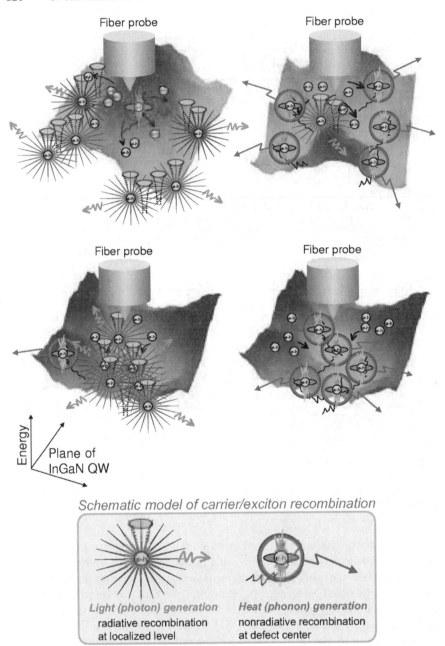

Fig. 41. Schematics of recombination model probed at positions of (**a**), (**b**), (**c**), and (**d**)

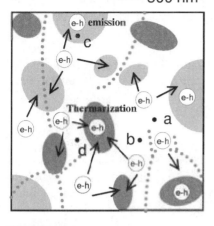

— 500 nm

Radiative recombination center

Nonradiative recombination center

→ Carriers and/or excitons transfer

········· High potential energy region

Fig. 42. Schematics of the different carrier dynamics as observed in Fig. 38 with its four studied positions of (**a**), (**b**), (**c**), and (**d**)

6   Conclusion

Materials parameters, such as bandgap energies, piezoelectric constants and alloy broadening have been summarized for In$_x$Ga$_{1-x}$N alloys, the general model of optical transitions has been described in In$_x$Ga$_{1-x}$N/GaN QW structures taking into account two major effects of localization and screening of piezoelectric fields.

The In$_{0.1}$Ga$_{0.9}$N well-width dependence on transient absorption revealed that the screening of internal electric fields plays an important role in the photoinduced change of absorption spectra in wider wells of 30 nm, 10 nm. However, this effect was less dominant for the well width of 5 nm, and was not detectable for the well width of 3 nm, resulting in the observation of photobleaching of localized tail states.

PL mapping with SNOM has revealed the dense distribution of island-like localized structures, the size of which ranges from 20 nm to 70 nm in a 3-nm thick In$_{0.2}$Ga$_{0.8}$N SQW structure emitting in the blue spectral region. Moreover, local diffusion, radiative and nonradiative processes have been identified at RT in this structure, by employing multimode SNOM where TRPL data are taken by both I–C mode and I mode. It was found that the probed area could be classified into four different regions whose dominating processes are 1) radiative recombination within a probing aperture, 2) nonradiative recombination within an aperture, 3) diffusion of photogenerated excitons/carriers out of an aperture resulting in localized luminescence and 4) the same diffusion process as 3), but resulting in nonradiative recombination.

We believe our experimental technique can be a powerful tool for any nanophotonic materials because of an applicability to study carrier/exciton dynamics where spatial and temporal dynamics have to be taken into account.

Acknowledgements

The authors would like to express their deep gratitude to Professor Shigeo Fujita for providing the opportunity for this work, his continuous guidance and encouragement, and also for his sincere attitude to education and research devoted just before his death. Special thanks are also due to Dr Mitsuru Funato, Dr Koichi Okamoto and Dr Ruggero Micheletto for their valuable discussion and comments.

References

1. T. Mukai, H. Narimatsu, S. Nakamura: Jpn. J. Appl. Phys. **37**, L479 (1998)
2. T. Mukai, K. Takekawa, S. Nakamura: Jpn. J. Appl. Phys. **37**, L839 (1998)
3. M. Yamada, Y. Narukawa, T. Mukai: Jpn. J. Appl. Phys. **41**, L246 (2002)
4. S. Nagahama, N. Iwasa, M. Senoh, T. Matsushita, Y. Sugimoto, H. Kiyoku, T. Kozaki, M. Sano, H. Matsumura, H. Umemoto, K. Chocho, T. Mukai: Jpn. J. Appl. Phys. **39**, L647 (2000)
5. S. Nagahama, T. Yanamoto, M. Sano, T. Mukai: Jpn. J. Appl. Phys: **40**, L785 (2001)
6. K. Iida, T. Kawashima, A. Miyazaki, H. Kasugai, S. Mishima, A. Honshio, Y Miyatake, M. Iwaya, S. Kamiyama, H. Amano, I. Akasaki: Jpn. J. Appl. Phys. **43**, L499 (2003)
7. B. Gil: *Low-Dimensional Nitride Semiconductors* (Oxford University Press, Oxford 2002)
8. S. Chichibu, T. Azuhata, T. Sota, S. Nakamura: Appl. Phys. Lett. **69**, 4188 (1996)
9. Y. Narukawa, Y. Kawakami, M. Funato, Sg. Fujita, S. Nakamura: Appl. Phys. Lett. **70**, 981 (1997)
10. Y. Narukawa, Y. Kawakami, Sz. Fujita, Sg. Fujita, S. Nakamura: Phys. Rev. B **56**, R10024 (1997)
11. A. Satake, Y. Masumoto, T. Miyajima, T. Asatsuma, F. Nakamura, M. Ikeda: Phys. Rev. B **57**, R2041 (1998)
12. Y.H. Cho, T.J. Schmidt, S. Bidnyk, G.H. Gainer, J.J. Song, S. Keller, U.K. Mishra, S.P. DenBaars: Phys. Rev. B **61**, 7571 (2000)
13. I. Ho, G.B. Stringfellow: Appl. Phys. Lett. **69**, 2701 (1996)
14. T. Matsuoka: Appl. Phys. Lett. **71**, 105 (1997)
15. A. Koukitu, N. Takahashi, T. Taki, H. Seki: Jpn. J. Appl. Phys. **35**, L673 (1996)
16. T. Okumura, M. Ishida, T. Kamikawa: Jpn. J. Appl. Phys. **39**, 1044 (2000)
17. C. Kisielowski, Z.L. Weber, S. Nakamura: Jpn. J. Appl. Phys. **36**, 6932(1997)

18. N.A. El-Masry, E.L. Piner, S.X. Liu, S.M. Bedair: Appl. Phys. Lett. **72**, 40 (1998)
19. L. Nistor, H. Bender, A. Vantomme, M.F. Wu, J. Van Landuyt, K.P. O'Donnell, R. Martin, K. Jacobs, I. Moerman: Appl. Phys. Lett. **7**, 507 (2000)
20. M. Strassburg, A. Hoffman, I.L. Krestnikov, N.N. Ledentsov: Phys. Stat. Sol (a) **183**, 99 (2001)
21. Y.T. Moon, D.J. Kim, J.S. Park, J.T. Oh, J.M. Lee, Y.W. Ok, H. Kim, S.J. Park: Appl. Phys. Lett. **79**, 599 (2001)
22. H. Gotoh, H. Ando, T. Takagahara, H. Kamada, A. Chavez-Pirson, J. Temmyo: Jpn. J. Appl. Phys. **36**, 4204 (1997)
23. Y. Narukawa, Y. Kawakami, Sg. Fujita, S. Nakamura: Phys. Rev. B **59**, 10283 (1999)
24. Y. Kawakami, Y. Narukawa, K. Omae, Sg. Fujita, S. Nakamura: Phys. Stat. Sol. (a) **178**, 331 (2000)
25. P. Lefebvre, J. Allegre, B. Gil, A. Kavokine, H. Mathieu, W. Kim, A. Salvador, A. Botchkarev, H. Morkoc: Phys. Rev. B **57**, R9447 (1998)
26. P. Lefebvre, T. Taliercio, A. Morel, J. Allegre, M. Gallart, B. Gil: Appl. Phys. Lett. **78**, 1538 (2001)
27. T. Takeuchi, S. Sota, M. Katsuragawa, M. Komori, H. Takeuchi, H. Amano, I. Akasaki: Jpn. J. Appl. Phys. **36**. L382 (1997)
28. J.S. Im, H. Kolmer, J. Off, A. Sohmer, F. Scolz, A. Hangleiter: Phys. Rev. B **55**, R9435 (1998)
29. P. Riblet, H. Hirayama, A. Kinoshita, A. Hirata, T. Sugano, Y. Aoyagi: Appl. Phys. Lett. **75**, 2241 (1999)
30. C. Wetzel, T. Takeuchi, H. Amano, I. Akasaki: Phys. Rev. B **62**, R13302 (2000)
31. T. Kuroda, A. Takeuchi, T. Sota: Appl. Phys. Lett. **76**, 3753 (2000)
32. C. Wetzel, T. Takeuchi, H. Amano, I. Akasaki: J. Appl. Phys. **85**, 3786 (1999)
33. Y. Kawakami, A. Kaneta, K. Omae, A. Shikanai, K. Okamoto, G. Marutsuki, Y. Narukawa, T. Mukai, Sg. Fujita: Phys. Stat. Sol. (b) **240**, 337 (2003)
34. K. Omae, Y. Kawakami, Sg. Fujita, M. Yamada, Y. Narukawa, T. Mukai: Phys. Rev. B **65**, 073308 (2002)
35. K. Omae, Y. Kawakami, Sg. Fujita, Y. Narukawa, Y. Watanabe, T. Mukai: Phys. Stat. Sol. (a) **190**, 93 (2002)
36. K. Omae, Y. Kawakami, Sg. Fujita, Y. Narukawa, T. Mukai: Phys. Rev. B **68**, 085303 (2003)
37. M. Nomura, M. Arita, Y. Arakawa, S. Ashihara, S. Kako, M. Nishioka, T. Shimura, K. Kuroda: J. Appl. Phys. **94**, 6468, (2003)
38. Y. Kawakami, Y. Narukawa, K. Omae, S. Nakamura, Sg. Fujita: Mater. Sci. Eng. B **82**, 188 (2001)
39. T. Izumi, Y. Narukawa, K. Okamoto, Sg. Fujita, S. Nakamura, J. Lumim. **87-89**, 1196 (2000)
40. A. Kaneta, K. Okamoto, Y. Kawakami, G. Marutsuki, Y. Nakagawa, G. Shinomiya, T. Mukai, Sg. Fujita: Phys. Stat. Sol. (b) **228**, 93 (2001)
41. K. Okamoto, J. Coi, Y. Kawakami, M. Terazima, T. Mukai, Sg. Fujita: Jpn. J. Appl. Phys. **43**, 839 (2004)
42. S. Chichibu, K. Wada, S. Nakamura: Appl. Phys. Lett. **71**, 2346 (1997)
43. X. Zhang, D.H. Rich, J.T. Kobayashi, N.T. Kobayashi, P.D. Dapkus: Appl. Phys. Lett. **73**, 1430 (1998)

44. P.A. Crowell, D.K. Young, S. Keller, E.L. Hu, D.D. Awschalom: Appl. Phys. Lett. **72**, 927 (1998)
45. A. Vertikov, M. Kuball, A.V. Nurmikko, Y. Chen, S.-Y. Wang: Appl. Phys. Lett. **72**, 2645 (1998)
46. A. Vertikov, A.V. Nurmikko, K. Doverspike, G. Bulman, J. Edmond: Appl. Phys. Lett, **73**, 493 (1998)
47. A. Vertikov, I. Ozden, A.V. Nurmikko: Appl. Phys. Lett. **74**, 850 (1999)
48. D.K. Young, M.P. Mack, A.C. Abare, M. Hansen, L.A. Coldren, S.P. Denbaars, E.L. Hu, D.D. Awschalom: Appl. Phys. Lett. **74**, 2349 (1999)
49. A. Kaneta, T. Izumi, K. Okamoto, Y. Kawakami, Sg. Fujita, Y. Narita, T. Inoue, T. Mukai: Jpn. J. Appl. Phys. **40**, 110 (2001)
50. A. Kaneta, K. Okamoto, Y. Kawakami, Sg. Fujita, G. Marutsuki, Y. Narukawa, T. Mukai: App. Phys. Lett. **81**, 4353 (2002)
51. J. Kudrna, P.G. Gucciardi, A. Vinattieri, M. Colocci, B. Damiliano, F. Semond, N. Grandjean, J. Massies: Phys. Stat. Sol. (a) **190**, 155 (2002)
52. A. Kaneta, T. Mutoh, Y. Kawakami, Sg. Fujita, G. Marutsuki, Y. Narukawa, T. Mukai: Appl. Phys. Lett. **83**, 3462 (2003)
53. R. Micheletto, N. Yoshimatsu, A. Kaneta, Y. Kawakami, Sg. Fujita: Appl. Surf. Sci. **229**, 338 (2004)
54. G. Marutsuki, Y. Narukawa, T. Mitani, T. Mukai, Gi. Shinomiya, A. Kaneta, Y. Kawakami, Sg. Fujita: Phys. Stat. Sol. (a) **192**, 110 (2002)
55. H. Itoh, S. Watanabe, M. Goto, N. Yamada, M. Misra, A.Y. Kim, S.A. Stockman: Jpn. J. Appl. Phys. **42**, L1244 (2003)
56. K. Osamura, K. Nakajima, Y. Murakami: Solid State Commun. **11**, 617 (1972)
57. K. Osamura, S. Naka, Y. Murakami: J. Appl. Phys. **46**, 3432 (1975)
58. C. Wetzel, T. Takeuchi, S. Yamaguchi, H. Katoh, H. Amano, I. Akasaki: Appl. Phys. Lett. **73**, 1994 (1998)
59. L. Bellaiche, T. Mattila, L.W. Wang, S.H. Wei, A. Zunger: Appl. Phys. Lett. **74**, 1842 (1999)
60. T. Inushima, V.V. Mamutin, V.A. Vekshin, S.V. Ivanov, T. Sakon, S. Motokawa: J. Cryst. Growth **227**, 481 (2001)
61. V.Y. Davydov, A.A. Klochikin, R.P. Sesyan, V.V. Emstev, S.V. Ivanov, F. Bechstedt, J. Furthmuller, H. Harima, A.V. Murdryi, J. Adrhold, O. Semchinova, J. Graul: Phys. Stat. Sol. (b) **229**, R1 (2002)
62. T. Matsuoka, H. Okamoto, M. Nakao, H. Harima, E. Kurimoto: Appl. Phys. Lett. **81**, 1246 (2002)
63. Y. Saito, H. Harima, E. Kueimoto, T. Yamaguchi, N. Teraguchi, A. Suzuki, T. Araki, Y. Nanishi: Phys. Stat. Sol. (b) **234**, 796 (2001)
64. M. Hori, K. Kano, T. Yamaguchi, Y. Saito, T. Araki, Y. Nanishi, N. Teraguchi, A. Suzuki: Phys. Stat. Sol. (b) **234**, 750 (2002)
65. J. Wu, W. Walukiewicz, K.M. Yu, J.W. Arger, III, E.E. Haller, H. Lu, W.J. Schaff, Y. Saito, Y. Nanishi: Appl. Phys. Lett. **80**, 3967 (2002)
66. Y. Nanishi, Y. Saito, T. Yamaguchi: Jpn. J. Appl. Phys. **42**, 2549 (2003)
67. I.M. Lifshitz: Adv. Phys. **13**, 483 (1965)
68. R. Zimmermann: J. Cryst. Growth **101**, 346 (1990)
69. V.W.L. Chin, T.L. Tansley, T. Osotchan: J. Appl. Phys. **75**, 7365 (1994)
70. R.B. Zetterstrom: J. Mater. Sci. **5**, 1102 (1970)
71. Y.C. Yeo, T.C. Chong, M.F. Li: J. Appl. Phys. **83**, 1429 (1998)
72. I. Vurgaftman, J.R. Meyer: J. Appl. Phys. **94**, 3675 (2003)

73. F. Bernardini, V. Fiorentini, D. Vanderbilt: Phys. Rev. B **56**, R10024 (1997)
74. A.D. Bykhovski, V.V. Kaminski, M.S. Shur, Q.C. Chen, M.A. Khan: Appl. Phys. Lett. **68**, 818 (1996)
75. S. Chichibu, T. Sota, K. Wada, O. Brandt, K.H. Ploog, S.P. DenBaars, S. Nakamura: Phys. Stat. Sol. (a) **183**, 91 (2001)
76. C. Gourdon, P. Lavallard: Phys. Stat. Sol. (b) **153**, 641 (1989)
77. Y. Kawakami: 'The optical properties of InGaN-based quantum wells and quantum dots', In: *Low-dimensional Nitride Semiconductors* ed. by B. Gill (Oxford University Press, Oxford 2002) pp. 233–255
78. S. Permogorov, A. Reznitskii, S. Verbin, G.O. Müller, P. Flögel, M. Nikiforov: Phys. Stat. Sol. (b) **113**, 589 (1982)
79. Y. Kawakami, M. Funato, Sz. Fujita, Sg. Fujita, Y. Yamada, Y. Masumoto: Phys. Rev. B **50**, 14655 (1994)
80. S.F. Chichibu, A.C. Abare, M.S. Minsky, S. Keller, S.B. Fleischer, J.E. Bowers, E. Hu, U.K. Mishra, L.A. Coldren, S.P. DenBaars, T. Sota: Appl. Phys. Lett. **73**, 2006 (1998)
81. E. Berkowicz, D. Gershoni, G. Bahir, E. Lakin, D. Shilo, E. Zolotoyabko, A.C. Abare, S.P. Denbaars, L.A. Coldren: Phys. Rev. B **61**, 10994 (2000)
82. D. Cherns, S.J. Heniey, F.A. Ponce: Appl. Phys. Lett. **78**, 2691 (2001)

Quantum Theory of Radiation in Optical Near Field Based on Quantization of Evanescent Electromagnetic Waves Using Detector Mode

Tetsuya Inoue and Hirokazu Hori

1 Introduction

After the rapid development of scanning near-field optical microscopes, near-field optics and the related techniques have been extended to manipulation and fabrication of nanometer-sized devices and, at present, have become an indispensable field of study in mesoscopic physics and nanotechnology with optical fields [1]. In contrast to the rapid progress in experimental works, there still remain a number of basic problems in the theoretical treatment of optical near fields. This is because it involves the difficult problem of electromagnetic interactions in its explicit form, which is still one of the most challenging problems of modern physics after the developments of quantum electrodynamics.

In this chapter, we investigate a theoretical treatment of optical near-field interactions based on the angular-spectrum representation of scattered fields developed for half-space problems [2] and field quantization based on the detector modes [3]. The purpose of the optical near-field theory developed in this work is not to calculate the optical fields consistent with the environmental material excitation but to evaluate the processes of excitation transfer and dissipation in subwavelength-sized material systems so as to provide the basis to develop optoelectronic devices that exhibit characteristic functions in optical near field. Strictly speaking, an optical near-field problem corresponds to an approximation of the optical problem where the whole system involves optical and electronic interactions between a number of optical and electronic components including optical sources and sinks. However, we can find a well-established meaning of its own even in a local subsystem of subwavelength size, provided that the optical interactions relevant to the function of the local subsystem have a characteristic property that is qualitatively different from those of the entire optical system except for the local subsystem considered. One of the purposes of the optical near-field theory is to provide a criterion to identify whether the local subsystem is separable or inseparable in its function from the entire optical system. In doing this, we should be careful about the energy transfer and dissipation that characterize the function of both the local and entire optical systems [4].

The study of the energy dissipation through the electromagnetic field is also very important in the theoretical description of nanometer-sized elec-

tronic devices. These problems manifest themselves in Coulomb blockade and
light emission associated with electron tunneling in scanning tunneling microscopes, as well as the optical radiation from quantum dots and molecules
under the electromagnetic influence of the environmental system such as a
substrate. In particular, for the problem of radiation associated with electron
tunneling, we do not yet have any fully developed theory for the lack of the
well-established basis to treat electromagnetic fields and electronic behaviors
on an equal footing.

In this chapter we will develop the theoretical treatments of optical near
fields and optical near-field interactions with all these issues into our scope. To
this end we will study the half-space problems based on the angular-spectrum
representation of scattered fields, where we can make clear the energy transfer between interacting objects separated by an assumed planar boundary
on which the electromagnetic energy transport is evaluated in terms of the
Poynting vector. This implies that even in the optical near-field problems one
can introduce a clear definition of optical source and sink, which provides the
basis to investigate the signal transport and associated dissipation in general
nano-optics devices.

Theoretical treatments of the light-scattering problems in half-space provide us with the basis of understanding, description, and evaluation of optical
near-field phenomena and investigation of its applications to nano-optics devices, such as optical near-field microscopes and optoelectronic devices of
nanometer size. Studies of the radiation processes in half-space are also the
foundation of optical manipulation and optical control of electronic systems
of nanometer size that compose nano-optics devices and exert function in the
complex material environments. Furthermore, the optical near-field interactions between nanometer-sized objects, such as molecules, quantum dots, and
quantum wires, play an important role in the process of excitation transfer
or information transport in the electronic devices of nanometer size. In this
chapter, we will investigate the basic description of half-space scattering problems of optical fields, the quantum theory of radiation in the optical near-field
regime, and the nature of optical near-field interactions as the foundation of
nanometer-sized optoelectronic devices.

1.1   Half-Space Problems and Angular-Spectrum Representation

The scattering processes of optical waves in half-space involve a number of
interesting phenomena that have been investigated extensively and are still
issues of great interest in relation to optical near-field interactions and nano-
optics devices. In general, an optical half-space problem is considered under
a system with a planar dielectric boundary, where an incidence of homogeneous plane waves from the higher refractive index side with an angle beyond
the critical angle of total internal reflection produces evanescent waves in the
lower-index side [5]. The evanescent wave is the most remarkable manifes-
tation of optical near field that propagates parallel to the boundary surface

exhibiting an exponential decay in the direction normal to the surface. Since the energy flux associated with the evanescent wave vanishes in the direction normal to the surface, the incident optical energy can not be transmitted into the other half-space and is totally reflected back to the dielectric medium of the source side. On the other hand, when a homogeneous plane wave is incident on the planar dielectric boundary from the lower refractive index side, the homogeneous wave is transmitted into the higher-index side at a propagation direction within the critical angle of total internal reflection. What is remarkable takes place when an evanescent wave is incident on the planar dielectric boundary from the half-space of lower refractive index; a homogeneous wave is transmitted into the higher-index side with a propagation direction out of the critical angle of total internal reflection. This process of excitation transfer via evanescent waves is the fundamental process of optical near-field interactions. One piece of the experimental evidence is given by the observation of radiation from an oscillating electric point dipole, such as an excited atom, placed in a subwavelength vicinity of a planar dielectric surface, where optical waves are excited in the dielectric medium regardless of the propagation direction within or out of the critical angle. The radiation field from the electric dipole involves both homogeneous and evanescent waves, when it is expanded in terms of the wavevector parallel to the boundary surface, so that the observed propagating waves in the directions out of the critical angle of total internal reflection can be attributed to the excitation transfer via evanescent waves. One of the remarkable features of this process is that the optical near-field interaction involves the additional paths of optical radiation into the medium that enhances the radiative decay of the oscillating dipole compared with that in free space. In the process of the optical near-field interaction, the conservation law holds for the pseudomomentum of the evanescent wave corresponding to the wavevector parallel to the boundary surface. In the near-field regime, the angular spectrum of the scattered field involves evanescent waves with wavevectors along the surface much larger than that of optical waves in free space. This results in the ultrahigh resolution of optical near-field microscopes and ultrahigh spatial controllability of optical near-field manipulation of subwavelength-sized matter.

As we have discussed above, the angular-spectrum representation provides us with the useful basis to study the optical near-field problems based on the half-space system. In this chapter, we will develop the theoretical study of the optical near-field interactions of excited electric and magnetic multipoles with evanescent waves based on an angular-spectrum representation of radiation fields [6]. We will show that the optical near-field interaction corresponding to the coupling to the evanescent waves can be evaluated in terms of the Poynting vectors corresponding to an overlap integral of evanescent waves over a planar surface lying between the interacting objects.

1.2 Quantization of Evanescent Electromagnetic Fields and Radiative Decay in Optical Near Field

The enhancement of radiation in optical near fields discussed in the above corresponds to a so-called cavity quantum electrodynamics (cavity QED) effect, which is a general term for enhancement and suppression of photon emission of excited electronic systems, such as atoms, molecules, and quantum dots, near matter studied on the basis of quantum theory of radiation [1,7–11]. The important issues are enhancement or reduction of spontaneous emission and the associated level shift of the radiating system due to variations of the environmental electromagnetic mode and to multiple interactions via the scattered electromagnetic field. Recently, these effects are utilized in controlling the motion and radiative properties of atoms and molecules especially by using high-Q optical resonators [12–15]. Cavity QED phenomena in a broad sense arise also in optical near-field regimes, or in optical near-field interactions of a radiating system with matter lying in its subwavelength vicinity. Extensive studies have been made on the radiation properties of atoms and molecules near a material surface [16–18] and also on the applications for atom manipulation [19–21]. Further interest in the near-field regime lies in photon-emission characteristics of mesoscopic electronic systems, such as quantum dots and wires, fabricated on a substrate and also in its observation process by means of optical near-field microscopy and spectroscopy. One can expect several interesting effects in the near-field regime that reflect the features of optical near field as an effective field or a coupled mode of electromagnetic field with matter. That is, the dispersion relations and polarization characteristics of optical near field deviate from those of photons in vacuum. Although an optical near-field interaction, in general, sustains a relatively short period compared with those with high-Q cavities it still exerts a considerable effect because of the high field intensity of the spatially localized field. For further study in this direction, it is important to develop the quantum-mechanical treatment of optical near field.

In a full quantum treatment of radiation of the atom or molecule, it is necessary to quantize the electromagnetic fields in half-space. The quantization of electromagnetic fields in half-space carried out by Carniglia and Mandel introduced the so-called triplet mode [22]. The triplet mode is described by a set of incident, reflected, and transmitted waves connected via Fresnel relations at the planar boundary under consideration. The complete set of triplet modes involved the evanescent photon. The triplet mode involving single incident wave serves as a convenient basis for the theoretical treatment of photon-absorption processes near a dielectric surface, where a single light source placed in far field may be assumed in a practical setup. In contrast, in accounting for the photon-emission processes near the boundary, the triplet mode should be related to a correlation measurement of photons by using two photodetectors coupled to each of the outgoing waves involved in the triplet. On the other hand, when we study photon-emission processes near a dielectric surface, we usually consider a practical setup with a photon-counting scheme by using an independent photodetector placed in each of the half-spaces separated by the boundary, so that we may consider a single outgoing wave as the final state of the radiation process. In this case, the so-called detector-mode function including a single outgoing wave serves as the convenient basis for theoretical analysis.

The radiative decay rate of atomic excitation has been studied both experimentally and theoretically for atomic and molecular dipoles put near a planar dielectric surface [10,23–25]. According to Fermi's golden rule the probability of a quantum-mechanical transition depends both on the transition matrix element and the density of the final state. The source of our interest in the cavity-QED problems lies in the controllability of the final-state density especially for the electromagnetic mode involved in the radiative transition considered. That is, the electromagnetic final state depends strongly on the scheme of our experiment. Therefore, in a practical analysis of experimental cavity-QED results, we should consider sources and sinks of photons as the reservoir being implemented outside of the photonic system (i.e., far-field observation), each of which couples to one of the incident and outgoing wave components belonging to the photonic mode under consideration.

In our context of radiation study, these are classified approximately into two categories: single-photon counting measurements and photon-correlation measurements. In the former only one of the photodetectors exclusively detects the single photon, but in the latter signals from several pairs of photodetectors exhibit correlation features reflecting coherence of radiation. In any case, the quantum-electrodynamic processes are evaluated in terms of external sources coupling to incident waves and external detectors coupling to outgoing waves being placed outside the isolated quantum system under consideration. Therefore, a careful consideration is required also on the role of sources and detectors for the practical near-field optical measurements concerned with both excitation and interaction of the local mode or its observation in far field, so that the entire process considered here exhibits the nature of an open system. Depending on our experimental schemes, we can select a convenient set of basis functions in describing the electromagnetic mode. Here, it is noted that, for the study of level shift of a radiative system, the important process lies rather in a closed system of multiple interactions between the radiative system and the environmental electromagnetic field.

1.3   Detector-Mode Description for Radiation Problem

In this chapter, we will study the field quantization based on the detector mode as the basis for theoretical analysis of the photon-emission process in optical near field [26]. We consider the radiative decay rate due to the electric dipole transition near a planar dielectric surface based on the novel 2nd quantization formalism developed on the basis of the detector mode. This gives us a straightforward evaluation of radiative decay rate in terms of the final-state density of photonic modes and provides a clear understanding of the meaning of detector mode as well as about the correspondence between classical and quantum descriptions of electric-dipole radiations of a two-level system in the near-field regime. The interaction between the atomic dipole and the evanescent photon corresponds to the tunneling process of energy flux due to the coupling to evanescent waves.

Reviewing the related works, a classical treatment of this problem for the magnetic and electric dipoles in half-space has been reported by Lukosz and Kunz [23–25], in which the boundary-value problem is solved using a combination of single-component magnetic and electric Hertz vectors to provide the total light emission intensity per unit time using Poynting's theorem. A semiclassical evaluation has been given by Wylie and Sipe for electric dipole radiation near a planar boundary based on general quantum-electrodynamic linear response theory [11,27]. We derive the full quantum treatment of electric quadrupole radiation near the planar boundary.

In this chapter, we also study the radiative decay rate due to the electric and magnetic multipole transition of arbitrary order near a planar dielectric surface, using the detector-mode functions. The optical near-field interaction process between the planar boundary and the atomic multipole transition is important, because this process corresponds to the light-scattering problem of an object with arbitrary shape [7,28,29].

1.4   Outline

Here, we present the outline of this chapter.

In Sect. 2, we present a brief review of the classical theory of dipole radiation in free space to introduce notations and basic formulations used in this chapter, where the nature of optical near fields is discussed in comparison with the asymptotic forms in the far-field regime. In order to evaluate the radiation decay rate of an oscillating dipole, the total radiation power is evaluated in terms of the Poynting vector.

In Sect. 3, we introduce the angular-spectrum representation of electromagnetic fields and calculate the total radiation power per second from an oscillating electric dipole. The angular-spectrum representation of scattered fields corresponds to a momentum-space expansion in a series of plane waves with complex wave number, which provides the convenient basis for half-space problems. In fact, one can separate near-field components with short penetration depth from propagating waves mediating long-ranged interaction. One can also study optical near-field interactions in terms of momentum conservation parallel to the boundary plane as well as of energy transfer through the boundary plane.

In Sect. 4, we study the classical theory of radiative decay of an oscillating electric dipole moment near a planar dielectric surface. We introduce a general treatment of half-space problems and discuss the fundamental processes involved in half-space problems, where the scattering processes are classified into three charactaristic categories, one of which is discussed in terms of excitation tunneling via evanescent waves. Then we will proceed to theoretical evaluation of radiative decay of an oscillating electric dipole on the basis of the angular-spectrum representation.

In Sect. 5, we study the field quantization based on the detector mode as the basis of a theoretical analysis of the photon-emission process in optical near field. The detector-mode-based theory gives us a straightforward evaluation of radiative decay rate in terms of the final-state density of the photonic mode and provides a clear understanding of the meaning of the detector mode as well as the correspondence between classical and quantum descriptions of electric and magnetic multipoles radiations of a two-level system in the near-field regime. We also study how the spontaneous-emission probability depends on the atomic excited state.

In Sect. 6, we extend our study to quantum optical theory of multipole radiation in optical near fields based on the detector mode and show that the multipole radiations are strongly enhanced in optical near field. The study of optical multipole radiation in near field provides an important basis in considerations of nanometer-sized electronic devices in terms of optical near-field interactions.

In Sect. 7, we introduce the tunneling picture of optical near-field interactions based on the calculation of the Poynting vector of scattered fields, using the angular-spectrum representation of electromagnetic fields. It is shown that the energy transfer of the tunneling regime takes place only through the overlap integral of evanescent waves with the same penetration depth and pseudomomentum involved in the angular spectra of scattered fields of interacting objects. We will clarify the role of dissipation processes that actually determine the transport of electromagnetic excitation.

2   Classical Theory of Radiation from an Oscillating Electric Dipole in Free Space

In this section, we will present a brief review of the classical theory of radiation from an oscillating point dipole in free space (vacuum) to introduce notations and basic formulations used in the following sections of this chapter. The nature of optical near fields is discussed in comparison with the asymptotic forms of the dipole fields in the far-field regime. In order to evaluate the radiation decay rate of an oscillating dipole associated with single-photon emission, we calculate the total radiation power of the dipole fields in terms of the Poynting vector.

2.1   Dipole Radiation in Free Space

We will consider the radiation from an oscillating electric dipole moment $d(r,t)$ with frequency K distributed in an infinitesimal spatial domain D

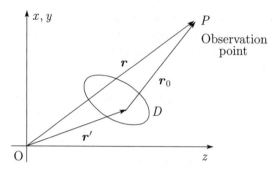

Fig. 1. Coordinate system employed: Light source is placed in a finite domain D, and radiation fields are evaluated at an observation point P

in vacumm with the coordinate system shown in Fig. 1. Throughout this chapter, we employ the unit in which the light velocity is taken to be unity, $c = 1$. We introduce the complex amplitude of the electric dipole, $d(r)$, as

$$d(r,t) = d(r)\exp(-\mathrm{i}Kt) + d^*(r)\exp(+\mathrm{i}Kt) , \tag{1}$$

where $d(r)$ is taken as a nonzero value in D, and zero outside D.

We consider vector potential $A^{(0)}(r,t)$ and scalar potential $\Phi^{(0)}(r,t)$ of monochromatic radiation fields and introduce complex amplitudes $A^{(0)}(r)$ and $\Phi^{(0)}(r)$ for positive-frequency Fourier components with frequency K. The complex amplitudes satisfy the Maxwell's equations

$$\left(\nabla^2 + K^2\right) A^{(0)}(r) = \frac{\mathrm{i}K}{\epsilon_0}d(r) , \tag{2}$$

$$\left(\nabla^2 + K^2\right) \Phi^{(0)}(r) = -\frac{1}{\epsilon_0}\rho(r) , \tag{3}$$

in Lorentz gauge

$$\nabla \cdot A^{(0)}(r) - \mathrm{i}K\Phi^{(0)}(r) = 0 , \tag{4}$$

where the charge density, $\rho(r)$, results from the distribution of the dipole as $\rho(r) = -\nabla \cdot d(r)$. The complex amplitudes $A^{(0)}(r)$ and $\Phi^{(0)}(r)$ outside the domain D are represented in terms of the (free-space) Green's function defined by

$$\left(\nabla^2 + K^2\right) G(r,r') = \delta(r - r') , \tag{5}$$

as

$$A^{(0)}(r) = \frac{\mathrm{i}K}{\epsilon_0} \int \mathrm{d}^3r' G(r,r')d(r') , \tag{6}$$

$$\Phi^{(0)}(r) = \frac{1}{\epsilon_0} \int \mathrm{d}^3r' \nabla\left[G(r,r')\right] \cdot d(r') . \tag{7}$$

Here, we have employed the relations $\rho(\boldsymbol{r}) = -\nabla \cdot \boldsymbol{d}(\boldsymbol{r})$ and $\nabla' G(\boldsymbol{r}, \boldsymbol{r}') = -\nabla G(\boldsymbol{r}, \boldsymbol{r}')$. The explicit form of G is given by

$$G(\boldsymbol{r}, \boldsymbol{r}') = -\frac{iK}{4\pi} h_0^{(1)}(K|\boldsymbol{r} - \boldsymbol{r}'|), \tag{8}$$

with the zeroth-order spherical Hankel function of the first kind,

$$h_0^{(1)}(Kr) = \frac{1}{iKr} \exp(iKr). \tag{9}$$

It is noted that when ∇ operates on an arbitrary scalar function $\chi(r)$, i.e., $\chi(r)$ depends only on the argument r in the spherical coordinates (r, θ, ϕ), one can use the relation,

$$\nabla \chi(r) = \hat{\boldsymbol{r}} \frac{d}{dr} \chi(r). \tag{10}$$

We can utilize the following recursion relation for spherical Hankel functions;

$$h_{\ell+1}^{(1)}(\rho) = -\frac{d}{d\rho} h_\ell^{(1)}(\rho) + \frac{\ell}{\rho} h_\ell^{(1)}(\rho) = -\rho^\ell \frac{d}{d\rho} \left(\frac{h_\ell^{(1)}(\rho)}{\rho^\ell} \right). \tag{11}$$

Substituting (8) into (6) and (7) with (10) and (11), we obtain the explicit forms of the complex amplitudes as

$$\boldsymbol{A}^{(0)}(\boldsymbol{r}) = \left(\frac{K^2}{4\pi\epsilon_0} \right) \int d^3 r' h_0^{(1)}(Kr_0) \boldsymbol{d}(\boldsymbol{r}'), \tag{12}$$

$$\Phi^{(0)}(\boldsymbol{r}) = \left(\frac{iK^2}{4\pi\epsilon_0} \right) \int d^3 r' h_1^{(1)}(Kr_0) \hat{\boldsymbol{r}}_0 \cdot \boldsymbol{d}(\boldsymbol{r}'), \tag{13}$$

where we have introduced the directional unit vector $\hat{\boldsymbol{r}}_0 = \boldsymbol{r}_0 / r_0$ corresponding to $\boldsymbol{r}_0 = \boldsymbol{r} - \boldsymbol{r}'$.

The complex amplitudes of electric and magnetic fields, $\boldsymbol{E}^{(0)}(\boldsymbol{r})$ and $\boldsymbol{B}^{(0)}(\boldsymbol{r})$, are given, respectively, by

$$\boldsymbol{E}^{(0)}(\boldsymbol{r}) = iK\boldsymbol{A}^{(0)}(\boldsymbol{r}) - \nabla \Phi^{(0)}(\boldsymbol{r}), \tag{14}$$

$$\boldsymbol{B}^{(0)}(\boldsymbol{r}) = \nabla \times \boldsymbol{A}^{(0)}(\boldsymbol{r}). \tag{15}$$

Substituting (12) and (13) into (14) and (15) and using (10) and (11), we obtain the explicit forms of electromagnetic amplitudes of electric dipole radiation as follows;

$$\boldsymbol{E}^{(0)}(\boldsymbol{r}) = \left(\frac{iK^3}{4\pi\epsilon_0} \right) \int d^3 r'_s \left\{ -\frac{1}{3} \left[\tilde{\boldsymbol{d}}(\boldsymbol{r}'_s) - 3(\hat{\boldsymbol{r}}_0 \cdot \tilde{\boldsymbol{d}}(\boldsymbol{r}'_s)) \hat{\boldsymbol{r}}_0 \right] h_2^{(1)}(Kr_0) \right.$$
$$\left. + \frac{2}{3} \tilde{\boldsymbol{d}}(\boldsymbol{r}'_s) h_0^{(1)}(Kr_0) \right\}, \tag{16}$$

$$\boldsymbol{B}^{(0)}(\boldsymbol{r}) = \left(\frac{-K^3}{4\pi\epsilon_0} \right) \int d^3 r'_s h_1^{(1)}(Kr_0) \hat{\boldsymbol{r}}_0 \times \tilde{\boldsymbol{d}}(\boldsymbol{r}'_s). \tag{17}$$

Here, we have utilized the following variation of the recursion relation;

$$h_{\ell-1}^{(1)}(\rho) = \frac{\mathrm{d}}{\mathrm{d}\rho} h_\ell^{(1)}(\rho) + \frac{\ell+1}{\rho} h_\ell^{(1)}(\rho) . \tag{18}$$

We introduce $r = r_s + R$ and $r' = r'_s + R$, where R indicates the position vector of a point in D. The integration with respect to $\mathrm{d}^3 r'$ has been converted to that with respect to $\mathrm{d}^3 r'_s$, and we have written $\tilde{d}(r'_s) = d(r'_s + R)$, and $r_0 = r - r' = r_s - r'_s$.

In order to evaluate the radiation amplitudes from an oscillating point dipole placed at R, we can replace the dipole distribution by

$$\tilde{d}(r'_s) = d\delta(r'_s) \tag{19}$$

with delta function $\delta(r'_s)$. Then, we obtain the complex electromagnetic amplitudes for the monochromatic electric dipole radiation as

$$E^{(0)}(r)$$
$$= \left(\frac{\mathrm{i}K^3}{4\pi\epsilon_0}\right) \left\{ -\frac{1}{3}[d - 3(\hat{r}_0 \cdot d)\hat{r}_0] h_2^{(1)}(Kr_0) + \frac{2}{3} dh_0^{(1)}(Kr_0) \right\} , \tag{20}$$

$$B^{(0)}(r) = \left(\frac{-K^3}{4\pi\epsilon_0}\right) h_1^{(1)}(Kr_0)\hat{r}_0 \times d , \tag{21}$$

where $r_0 = r - R$.

In the conventional treatment of dipole radiation, an observation is assumed to be made by using a photodetector placed in the far-field region. In such a case where the far-field condition, $(Kr_0 \gg 1)$, holds, the asymptotic forms of $E^{(0)}(r)$ and $B^{(0)}(r)$ are available for evaluation of radiation intensity. The asymptotic forms of the spherical Hankel functions for large argument,

$$h_\ell^{(1)}(\rho) \sim \frac{1}{\rho} \exp\{\mathrm{i}[\rho - (\ell+1)\frac{\pi}{2}]\} \quad \text{for } \rho \gg 1 ,$$

provide the dipole radiation fields in the far-field regime, $Kr_0 \gg 1$, as

$$E^{(0)}(r) \sim \left(\frac{K^2}{4\pi\epsilon_0}\right) [d - (\hat{r}_0 \cdot d)\hat{r}_0] \frac{\exp(\mathrm{i}Kr_0)}{r_0} , \tag{22}$$

$$B^{(0)}(r) \sim \left(\frac{K^2}{4\pi\epsilon_0}\right) (\hat{r}_0 \times d) \frac{\exp(\mathrm{i}Kr_0)}{r_0} . \tag{23}$$

These asymptotic forms correspond to ordinary spherical waves.

On the other hand, when an observation is made near the point dipole, the near-field condition, $Kr_0 \ll 1$, leads to other asymptotic forms of $E^{(0)}(r)$ and $B^{(0)}(r)$ that are useful for understanding the nature of optical phenomena of the near-field regime. For small arguments $\rho \ll 1$, the ℓ-th-order spherical Bessel function $j_\ell(\rho)$ and Neumann function $n_\ell(\rho)$ are approximated, respectively, by

$$j_\ell(\rho) \sim \frac{\rho^\ell}{(2\ell+1)!!} , \quad n_\ell(\rho) \sim -\frac{(2\ell-1)!!}{\rho^{\ell+1}} \quad \text{for } \rho \ll 1 .$$

Applying these to the spherical Hankel function, $h_\ell^{(1)}(\rho) = j_\ell(\rho) + in_\ell(\rho)$, the dipole radiation fields in the near-field regime, $Kr_0 \ll 1$, are obtained as follows;

$$\boldsymbol{E}^{(0)}(\boldsymbol{r}) \sim \left(\frac{iK^3}{4\pi\epsilon_0}\right) \left\{ [\boldsymbol{d} - 3(\hat{\boldsymbol{r}}_0 \cdot \boldsymbol{d})\hat{\boldsymbol{r}}_0] \frac{i}{K^3 r_0^3} + \frac{2}{3}\boldsymbol{d} \right\} , \tag{24}$$

$$\boldsymbol{B}^{(0)}(\boldsymbol{r}) \sim \left(\frac{-K^3}{4\pi\epsilon_0}\right) (\hat{\boldsymbol{r}}_0 \times \boldsymbol{d}) \left(\frac{Kr_0}{3} - \frac{i}{K^2 r_0^2}\right) . \tag{25}$$

These results clearly show that one observes an extremely large field amplitude in optical near field of an oscillating point dipole. Here, it is noted that both the real and imaginary components involved in (24) and (25) play an important role in optical near-field interactions. As we will see later, products of the complex amplitudes, such as $\boldsymbol{E}^{(0)} \cdot \boldsymbol{E}^{(0)*}$ and $\boldsymbol{E}^{(0)} \times \boldsymbol{B}^{(0)*}$, contribute to power density and energy flow in the optical near-field regime.

2.2   Total Radiation Intensity in Free Space

The electromagnetic energy flux is given in terms of the Poynting vector defined by

$$\boldsymbol{P}(\boldsymbol{r},t) = \epsilon_0 \boldsymbol{E}(\boldsymbol{r},t) \times \boldsymbol{B}(\boldsymbol{r},t) , \tag{26}$$

where $\boldsymbol{E}(\boldsymbol{r},t)$ and $\boldsymbol{B}(\boldsymbol{r},t)$ are electric and magnetic fields, respectively, and the unit $\epsilon_0\mu_0 = 1$ ($c = 1$) is employed. The temporal average of the total radiation power per second, I, from the oscillating electric dipole is given as the temporal average of the outgoing Poynting vector through a closed surface ∂D enclosing the source in the domain D:

$$I = \oint_{\partial D} \langle \boldsymbol{P}(\boldsymbol{r},t) \rangle \cdot \hat{\boldsymbol{r}}_0 d\sigma , \tag{27}$$

where $d\sigma$ is an infinitesimal area on ∂D and $\hat{\boldsymbol{r}}_0$ is its unit normal vector (see Fig. 2).

The temporal average of the Poynting vector $\langle \boldsymbol{P}(\boldsymbol{r},t) \rangle$ is represented in terms of the complex amplitudes of dipole fields, $\boldsymbol{E}(\boldsymbol{r})$ and $\boldsymbol{B}(\boldsymbol{r})$, as

$$\langle \boldsymbol{P}(\boldsymbol{r},t) \rangle = 2\epsilon_0 \Re e \left\{ \boldsymbol{E}(\boldsymbol{r}) \times \boldsymbol{B}^*(\boldsymbol{r}) \right\} . \tag{28}$$

For the case of the point dipole, it is convenient to take ∂D as a sphere of radius r_0 centered at the dipole. Employing polar coordinates (θ, ϕ) for representation of the unit vector $\hat{\boldsymbol{r}}_0$,

$$\hat{\boldsymbol{r}}_0 = (\sin\theta\cos\phi, \sin\theta\sin\phi, \cos\theta) , \tag{29}$$

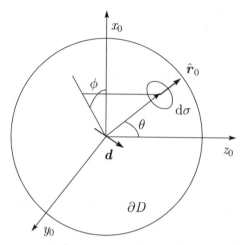

Fig. 2. The total radiation power per second, I, from an oscillating electric dipole is evaluated by integration of the temporal average of the Poynting vector outgoing through a closed surface ∂D enclosing the source in the domain D. $d\sigma$ is the infinitesimal area on ∂D and $\hat{n}_0$ is the corresponding unit normal vector

where $d\sigma = r_0^2 \sin\theta d\theta d\phi$, we obtain the total radiation power per second as

$$I^{(0)} = 2\epsilon_0 r_0^2 \int_0^\pi \sin\theta d\theta \int_0^{2\pi} d\phi \ \Re e \left\{ \boldsymbol{E}^{(0)}(\boldsymbol{r}) \times \boldsymbol{B}^{(0)*}(\boldsymbol{r}) \right\} \cdot \hat{\boldsymbol{r}}_0 \ . \tag{30}$$

Substituting (20) and (21) into (30), we obtain the total radiation power per second as

$$I^{(0)} = \left(\frac{K^4}{8\pi^2\epsilon_0} \right) \int_0^\pi \sin\theta d\theta \int_0^{2\pi} d\phi \left[|\boldsymbol{d}|^2 - (\boldsymbol{d}^* \cdot \hat{\boldsymbol{r}}_0)(\hat{\boldsymbol{r}}_0 \cdot \boldsymbol{d}) \right] \ , \tag{31}$$

where the following relation is utilized;

$$\frac{1}{3}\Im m \left\{ \left[2h_0^{(1)}(\rho) - h_2^{(1)}(\rho) \right] h_1^{(1)*}(\rho) \right\} = \Im m \left\{ \frac{\partial h_0^{(1)}(\rho)}{\partial \rho} h_0^{(1)*}(\rho) \right\} \ , \tag{32}$$

which is derived from the recursion relations in (11) and (18). The radiation power $I^{(0)}$ derived from $\boldsymbol{E}^{(0)}(\boldsymbol{r})$ and $\boldsymbol{B}^{(0)}(\boldsymbol{r})$ is independent of the distance r_0 between the point dipole and the observation point. It is confirmed that the same results can be obtained by using the asymptotic forms given in (22) and (23) for the far-field condition, as well as (24) and (25) for the near-field regime. Taking the dipole orientation in the z-axis, for further convenience,

$$\boldsymbol{d} = |\boldsymbol{d}| \, \hat{\boldsymbol{e}}_z \ . \tag{33}$$

The result is rewritten as

$$I^{(0)} = \left(\frac{K^4|\boldsymbol{d}|^2}{8\pi^2\epsilon_0} \right) \int_0^\pi \sin\theta d\theta \int_0^{2\pi} d\phi \left(1 - \cos^2\theta \right) = \frac{K^4|\boldsymbol{d}|^2}{3\pi\epsilon_0} \ , \tag{34}$$

which is, as expected, independent of the direction of the electric dipole $\boldsymbol{d}$ due to spatial isotropy.

Here we proceed to the semiclassical theory of electric dipole radiation and consider the probability Γ of an electronic transition in the optical source between quantum-mechanical states $|\varphi_i\rangle$ and $|\varphi_f\rangle$ resulting in single-photon emission. Then, in order to evaluate the transition probability Γ, we can replace the classical electric dipole moment in (34) by the corresponding sum of the dipole transition matrix elements, $\boldsymbol{d}_{fi} = \langle\varphi_f|\boldsymbol{d}|\varphi_i\rangle$ over all the possible final electronic states of the transition as

$$|\boldsymbol{d}|^2 \to \sum_f |\boldsymbol{d}_{fi}|^2 \ .$$

The total radiation power per second, $I^{(0)}$, corresponding to the single-photon emission is then translated to the probability of photon emission $\Gamma^{(0)}$ when it is normalized by the photon energy $\hbar K$ as $\Gamma^{(0)} = I^{(0)}/(\hbar K)$, with photon energy as

$$\Gamma^{(0)} = \left(\frac{K^3}{3\pi\hbar\epsilon_0}\right) \sum_f |\boldsymbol{d}_{fi}|^2 \ . \tag{35}$$

We will later compare this result with those in half-space problems and investigate the modification of radiation properties due to the presence of a material surface nearby an oscillating electric dipole.

3   Classical Theory of Radiation Based on Angular-Spectrum Representation

In this section, we will study the angular-spectrum representation of scattered electromagnetic fields, which is very useful for description of optical near-field problems. Angular-spectrum representation corresponds to a momentum-space expansion of arbitrary scattered fields with respect to an assumed planar boundary in a series of plane waves with complex wave number. Therefore, it provides one of the most convenient bases for half-space problems that we encounter in near-field optics. The imaginary part of the wave number represents an exponential decay of the wave function in the corresponding direction that is normal to the assumed boundary plane. Such decaying plane waves are referred to as evanescent waves. Therefore, one can separate clearly near-field components with short penetration depth from propagating waves mediating long-ranged interaction. One can also study optical near-field interactions in terms of momentum conservation parallel to the boundary plane as well as of energy transfer through the boundary plane. These properties are, in general, the most important ones in the considerations of resolution and sensitivity related to optical near-field microscopy. In this study, we will focus on radiation properties of optically excited electronic

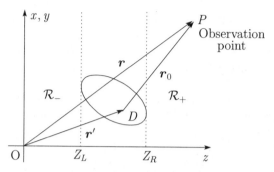

Fig. 3. Scattered and assumed planar boundaries as the basis of the angular-spectrum representation of scattered optical fields. The source of scattered fields lies in the domain D, and the right half-space separated by the planar boundary is referred to as $\mathcal{R}_+$ and left half-space as $\mathcal{R}_-$

system, in which angular-spectrum representation provides a number of advantageous properties related to descriptions of short penetration depth and pseudomomentum conservation in optical near-field interactions.

In the following, we introduce the angular-spectrum representation of electromagnetic fields and calculate the total radiation power per second from an oscillating electric dipole based on the angular-spectrum representation of electromagnetic fields.

3.1   Angular-Spectrum Representation

We consider an electromagnetic scattering problem with a pair of planar boundaries, regardless of whether the boundary planes correspond to real boundaries formed by different material media or these are just assumed. We assume that the source of scattered fields lies in the domain D, and the right half-space separated by the planar boundary is referred to as $\mathcal{R}_+$ ($Z_\mathrm{R} < z$), and left half-space as $\mathcal{R}_-$ ($z < Z_\mathrm{L}$). It is noted that the observation point of scattered fields lies in $z > z'$ in $\mathcal{R}_+$, and $z < z'$ in $\mathcal{R}_-$.

The angular-spectrum representation of scattered electromagnetic fields is obtained from that of outgoing scalar spherical wave $h_0^{(1)}(K|\boldsymbol{r} - \boldsymbol{r}'|)$, given by

$$h_0^{(1)}(K|\boldsymbol{r} - \boldsymbol{r}'|) = \frac{1}{2\pi} \int\!\!\int_{-\infty}^{+\infty} \mathrm{d}s_x \mathrm{d}s_y \frac{1}{s_z} \exp\left[\mathrm{i}K\hat{\boldsymbol{s}}^{(\pm)} \cdot (\boldsymbol{r} - \boldsymbol{r}')\right] , \qquad (36)$$

where $\hat{\boldsymbol{s}}^{(\pm)}$ stands for unit wavevector with Cartesian components

$$\hat{\boldsymbol{s}}^{(\pm)} = (s_x, s_y, \pm s_z) , \qquad (37)$$

which satisfies the dispersion relation of optical wave, $s_x^2 + s_y^2 + s_z^2 = 1$, and produces the solid angle $\mathrm{d}s_x \mathrm{d}s_y / s_z$ in the direction of the unit vector $\hat{\boldsymbol{s}}^{(\pm)}$ (see Fig. 4).

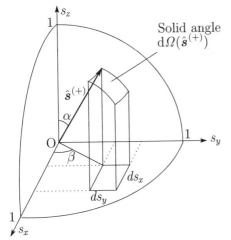

Fig. 4. The solid angle $d\Omega(\hat{s}^{(\pm)})$. Coordinate system corresponding to that employed for angular-spectrum representation of outgoing spherical waves. For real values of s_z, $\hat{s}^{(\pm)} = (\sin\alpha\cos\beta, \sin\alpha\sin\beta, \pm\cos\alpha)$ indicates the unit wavevector of the plane wave involved in the angular spectrum. Infinitesimal solid angle corresponds to the indicated area on the unit circle satisfies the relation $d\Omega(\hat{s}^{(\pm)})\cos\alpha = ds_x ds_y$, so that it is described as $d\Omega(\hat{s}^{(\pm)}) = ds_x ds_y / s_z$

The s_x and s_y are always real, but s_z are either real or pure imaginary corresponding, respectively, to homogeneous wave or evanescent wave. With $s_\parallel = \sqrt{s_x^2 + s_y^2}$, s_z is represented by

$$s_z = \begin{cases} \sqrt{1 - s_\parallel^2} & \text{for } 0 \le s_\parallel < 1, \\ i\sqrt{s_\parallel^2 - 1} & \text{for } 1 \le s_\parallel < +\infty. \end{cases} \tag{38}$$

One should take the unit wavevector $\hat{s}^{(+)}$ for $z > z'$, and $\hat{s}^{(-)}$ for $z < z'$. For real values of s_z, $\exp\left[iK\hat{s}^{(\pm)} \cdot (\boldsymbol{r} - \boldsymbol{r}')\right]$ represents a homogeneous outgoing plane wave from a point $\boldsymbol{r}'$ on the boundary plane. On the other hand, for pure imaginary values of s_z, $\exp\left[\mp K|s_z| \cdot z\right]\exp\left[iK\hat{s}_\parallel \cdot (\boldsymbol{r}_\parallel - \boldsymbol{r}'_\parallel)\right]$ represents an evanescent wave propagating parallel to the boundary plane showing an exponential decay with increase in distance from the boundary. In other words, as is widely recognized, far-field behavior and near-field behavior of scattered fields are dominated, respectively, by homogeneous plane waves and evanescent waves involved in the angular-spectrum representation of the fields.

Figure 5 shows the s_x–s_y cross section of the dispersion relation, $(s_x^2 + s_y^2 + s_z^2 = 1)$. The projections of wavevector, $\hat{s}_\parallel$ lying inside and outside the unit circle correspond, respectively, to homogeneous and evanescent waves. Figures 6a and b show the cross section of the dispersion relation in the com-

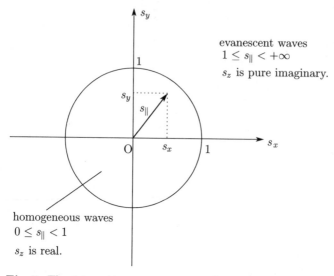

Fig. 5. The integration area corresponds to s_x and s_y. The inside of unit sphere $(0 \leq s_\| < 1)$ corresponds to the homogeneous waves. The outside of unit sphere $(1 \leq s_\| < +\infty)$ corresponds to the evanescent waves

plex s_z plane, where the circle with $0 \leq s_\| < 1$, therefore, with real colatitude α, corresponds to homogeneous waves with real values of s_z, and the hyperbola with $1 \leq s_\| < +\infty$, therefore, with complex colatitude α, represents evanescent waves with pure imaginary values of s_z. The complex argument of the integration involved in the angular-spectrum representation sweeps the entire surface of the dispersion relation. It should be noted that, for the case of an optical point source, the amplitudes of wave components involved in the angular-spectrum representation, or the angular spectrum of scattered fields, depends on the distance between the point source and the assumed boundary plane. In the near-field regime, according to the decrease in the distance, the amplitudes of evanescent waves with larger $s_\|$ and shorter penetration depth increase in the angular spectrum. This is the most significant property of near-field observations of electromagnetic radiation.

3.2   Angular-Spectrum Representation of Scattered Electromagnetic Fields

For the complex amplitudes of vector and scalar potentials corresponding to dipole radiation, the angular-spectrum representation is derived in each of the half-spaces $\mathcal{R}_\pm$ by substituting (36) into (6) and (7);

$$
\begin{aligned}
&\boldsymbol{A}^{(0)}(\boldsymbol{r}) \\
&= \left(\frac{K^2}{8\pi^2\epsilon_0}\right) \int\!\!\int_{-\infty}^{+\infty} ds_x ds_y \frac{1}{s_z} \tilde{\boldsymbol{d}}(K\hat{\boldsymbol{s}}^{(\pm)}) \exp\left[iK\hat{\boldsymbol{s}}^{(\pm)} \cdot (\boldsymbol{r} - \boldsymbol{R})\right] , \quad (39)
\end{aligned}
$$

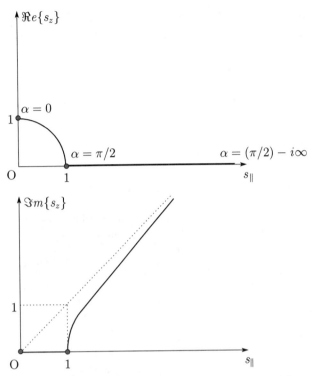

Fig. 6. (a) The *curve* corresponds to the real part of the z-element of $\hat{s}^{(+)}$, $\Re e\{s_z\}$. The *curve* meets the circle with radius 1 for the homogeneous mode $0 \leq s_\parallel < 1$. (b) The *curve* corresponds to the imaginal part of the z-element of $\hat{s}^{(+)}$, $\Im m\{s_z\}$. The *curve* meets the hyperbolic curve for the evanescent mode $1 \leq s_\parallel < +\infty$

$$
\Phi^{(0)}(\boldsymbol{r})
$$
$$
= \left(\frac{K^2}{8\pi^2\epsilon_0}\right) \int\!\!\int_{-\infty}^{+\infty} \mathrm{d}s_x \mathrm{d}s_y
$$
$$
\times \frac{1}{s_z}\left[\hat{\boldsymbol{s}}^{(\pm)} \cdot \tilde{\boldsymbol{d}}(K\hat{\boldsymbol{s}}^{(\pm)})\right] \exp\left[\mathrm{i}K\hat{\boldsymbol{s}}^{(\pm)} \cdot (\boldsymbol{r} - \boldsymbol{R})\right] . \quad (40)
$$

We have introduced $\boldsymbol{r}' = \boldsymbol{r}'_{\mathrm{s}} + \boldsymbol{R}$, where $\boldsymbol{R}$ stands for the position vector of a point in D, and $\tilde{\boldsymbol{d}}(K\hat{\boldsymbol{s}}^{(\pm)})$ indicates the Fourier spectrum of $\tilde{\boldsymbol{d}}(\boldsymbol{r}'_{\mathrm{s}}) = \boldsymbol{d}(\boldsymbol{r}'_{\mathrm{s}} + \boldsymbol{R})$ defined by

$$
\tilde{\boldsymbol{d}}(K\hat{\boldsymbol{s}}^{(\pm)}) = \int \mathrm{d}^3 r'_{\mathrm{s}} \exp\left(-\mathrm{i}K\hat{\boldsymbol{s}}^{(\pm)} \cdot \boldsymbol{r}'_{\mathrm{s}}\right) \tilde{\boldsymbol{d}}(\boldsymbol{r}'_{\mathrm{s}}) . \quad (41)
$$

The unit wavevector $\hat{\boldsymbol{s}}^{(+)}$ should be taken for the obsevation at a point in $\mathcal{R}_+$, and $\hat{\boldsymbol{s}}^{(-)}$ in $\mathcal{R}_-$.

Substituting (39) and (40) into (14) and (15), we obtain the angular-spectrum representation of complex amplitudes, $\boldsymbol{E}^{(0)}$ and $\boldsymbol{B}^{(0)}$, as follows,

$$
\boldsymbol{E}^{(0)}(\boldsymbol{r}) = \left(\frac{iK^3}{8\pi^2\epsilon_0}\right) \int\int_{-\infty}^{+\infty} ds_x ds_y \frac{1}{s_z}
$$
$$
\times \left\{ \tilde{\boldsymbol{d}}(K\hat{\boldsymbol{s}}^{(\pm)}) - \left[\hat{\boldsymbol{s}}^{(\pm)} \cdot \tilde{\boldsymbol{d}}(K\hat{\boldsymbol{s}}^{(\pm)})\right] \hat{\boldsymbol{s}}^{(\pm)} \right\} \exp\left[iK\hat{\boldsymbol{s}}^{(\pm)} \cdot (\boldsymbol{r} - \boldsymbol{R})\right] , \quad (42)
$$

$$
\boldsymbol{B}^{(0)}(\boldsymbol{r}) = \left(\frac{iK^3}{8\pi^2\epsilon_0}\right) \int\int_{-\infty}^{+\infty} ds_x ds_y \frac{1}{s_z}
$$
$$
\times \left[\hat{\boldsymbol{s}}^{(\pm)} \times \tilde{\boldsymbol{d}}(K\hat{\boldsymbol{s}}^{(\pm)})\right] \exp\left[iK\hat{\boldsymbol{s}}^{(\pm)} \cdot (\boldsymbol{r} - \boldsymbol{R})\right] . \quad (43)
$$

Here, we restrict ourselves to considering a point dipole placed at $\boldsymbol{R}$ as described in (19) and reduce (41), using (19), into the following forms;

$$
\boldsymbol{E}^{(0)}(\boldsymbol{r}) = \left(\frac{iK^3}{8\pi^2\epsilon_0}\right) \int\int_{-\infty}^{+\infty} ds_x ds_y \frac{1}{s_z}
$$
$$
\times \left[\boldsymbol{d} - \left(\hat{\boldsymbol{s}}^{(\pm)} \cdot \boldsymbol{d}\right) \hat{\boldsymbol{s}}^{(\pm)}\right] \exp\left[iK\hat{\boldsymbol{s}}^{(\pm)} \cdot (\boldsymbol{r} - \boldsymbol{R})\right] , \quad (44)
$$

$$
\boldsymbol{B}^{(0)}(\boldsymbol{r}) = \left(\frac{iK^3}{8\pi^2\epsilon_0}\right) \int\int_{-\infty}^{+\infty} ds_x ds_y \frac{1}{s_z}
$$
$$
\times \left(\hat{\boldsymbol{s}}^{(\pm)} \times \boldsymbol{d}\right) \exp\left[iK\hat{\boldsymbol{s}}^{(\pm)} \cdot (\boldsymbol{r} - \boldsymbol{R})\right] . \quad (45)
$$

It is convenient to introduce a set of polarization vectors $\varepsilon(\hat{\boldsymbol{s}}^{(\pm)}, 1)$ and $\varepsilon(\hat{\boldsymbol{s}}^{(\pm)}, 2)$ with respect to the unit wavevector $\hat{\boldsymbol{s}}^{(\pm)}$ defined, respectively, by

$$
\hat{\varepsilon}(\hat{\boldsymbol{s}}^{(\pm)}, 1) = \hat{\varepsilon} , \quad (46)
$$
$$
\hat{\varepsilon}(\hat{\boldsymbol{s}}^{(\pm)}, 2) = -\hat{\boldsymbol{s}}^{(\pm)} \times \hat{\varepsilon} . \quad (47)
$$

Here, $\hat{\varepsilon}$ is a real unit vector lying in the $z = 0$ plane orthogonal to the unit wavevector $\hat{\boldsymbol{s}}^{(\pm)}$. The label μ in $\hat{\varepsilon}(\hat{\boldsymbol{s}}^{(\pm)}, \mu)$ specifies the state of polarization so that $\mu = 1$ and 2 corresponds, respectively, to TE and TM waves. For example, when we consider a set of orthogonal unit vectors $(\hat{\boldsymbol{e}}_s, \hat{\boldsymbol{e}}_\alpha, \hat{\boldsymbol{e}}_\beta)$ as the basis of spherical coordinates and take $\hat{\boldsymbol{s}}^{(+)}$ as $\hat{\boldsymbol{e}}_s$, the polarization vectors $\hat{\varepsilon}(\hat{\boldsymbol{s}}^{(+)}, 1)$ and $\hat{\varepsilon}(\hat{\boldsymbol{s}}^{(+)}, 2)$ correspond, respectively, to $\hat{\boldsymbol{e}}_\beta$ and $\hat{\boldsymbol{e}}_\alpha$. Figure 7 shows how we define polarizations with respect to the unit wavevector $\hat{\boldsymbol{s}}^{(+)}$.

With the Cartesian coordinates shown in Fig. 7, TE- and TM-polarization vectors are represented, respectively, by

$$
\hat{\varepsilon}(\hat{\boldsymbol{s}}^{(\pm)}, 1) = (-\frac{s_y}{s_\|}, \frac{s_x}{s_\|}, 0) , \quad (48)
$$

$$
\hat{\varepsilon}(\hat{\boldsymbol{s}}^{(\pm)}, 2) = (\pm\frac{s_z s_x}{s_\|}, \pm\frac{s_y s_z}{s_\|}, -s_\|) . \quad (49)
$$

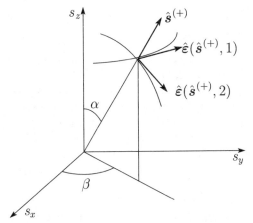

Fig. 7. The polarization vectors, $\hat{\varepsilon}(\hat{s}^{(+)}, 1)$ and $\hat{\varepsilon}(\hat{s}^{(+)}, 2)$, corresponding, respectively, to TE and TM waves, defined with respect to the unit wavevector $\hat{s}^{(+)}$

The TE-polarization vector $\hat{\varepsilon}(\hat{s}^{(\pm)}, 1)$ is always a real unit vector, since it does not involve the complex component of wavevector s_z. On the other hand, $\hat{\varepsilon}(\hat{s}^{(\pm)}, 2)$ is a real unit vector for $0 \le s_\| < 1$ and a complex vector for $1 < s_\| \le +\infty$. The unit wavevector and the polarization vectors satisfy the following orthogonality relations;

$$\hat{\varepsilon}(\hat{s}^{(\pm)}, \mu) \cdot \hat{\varepsilon}(\hat{s}^{(\pm)}, \mu') = \delta_{\mu\mu'} , \tag{50}$$

$$\hat{\varepsilon}(\hat{s}^{(\pm)}, \mu) \cdot \hat{s}^{(\pm)} = 0 , \tag{51}$$

where μ and μ' take values 1 and 2. Regarding these orthogonality relations, we can expand the electric dipole moment $\boldsymbol{d}$ as

$$\boldsymbol{d} = \hat{s}^{(\pm)} \left(\hat{s}^{(\pm)} \cdot \boldsymbol{d} \right) + \sum_{\mu=1}^{2} \hat{\varepsilon}(\hat{s}^{(\pm)}, \mu) \left[\hat{\varepsilon}(\hat{s}^{(\pm)}, \mu) \cdot \boldsymbol{d} \right] . \tag{52}$$

This transforms (44) and (45) into the following forms;

$$\boldsymbol{E}^{(0)}(\boldsymbol{r}) = \left(\frac{iK^3}{8\pi^2\epsilon_0} \right) \sum_{\mu=1}^{2} \int\!\!\int_{-\infty}^{+\infty} ds_x ds_y \frac{1}{s_z}$$
$$\times \left[\hat{\varepsilon}(\hat{s}^{(\pm)}, \mu) \cdot \boldsymbol{d} \right] \hat{\varepsilon}(\hat{s}^{(\pm)}, \mu) \exp\left[iK\hat{s}^{(\pm)} \cdot (\boldsymbol{r} - \boldsymbol{R}) \right] , \tag{53}$$

$$\boldsymbol{B}^{(0)}(\boldsymbol{r}) = \left(\frac{iK^3}{8\pi^2\epsilon_0} \right) \sum_{\mu=1}^{2} \int\!\!\int_{-\infty}^{+\infty} ds_x ds_y \frac{1}{s_z}$$
$$\times (-1)^\mu \left[\hat{\varepsilon}(\hat{s}^{(\pm)}, \mu) \cdot \boldsymbol{d} \right] \hat{\varepsilon}(\hat{s}^{(\pm)}, \bar{\mu}) \exp\left[iK\hat{s}^{(\pm)} \cdot (\boldsymbol{r} - \boldsymbol{R}) \right] , \tag{54}$$

where $\bar{\mu}$ stands for the interchanged suffix μ, i.e., $\bar{\mu} = 2$ for $\mu = 1$, and vice versa.

3.3   Angular Spectrum of Dipole Radiation Fields in Optical Near-Field Regime

Here, it is especially instructive to investigate the angular spectrum of the optical field due to electric-dipole radiation in the optical near-field regime. The dipole field invloves both homogeneous and evanescent waves in its angular spectrum. When the angular-spectrum representation is made with respect to an assumed boundary plane placed at a subwavelength distance z_0 from the electric dipole, the angular spectrum involves evanescent waves of large amplitude. This implies that the optical near-field interactions of the electric dipole with a material object placed at the distance z_0 are dominated by interactions via evanescent waves corresponding to the large amplitude components in the angular spectrum. We will actually evaluate the angular spectrum of dipole radiation fields in order to see the shape and magnitude of the angular spectrum depending on the distance z_0. This provides the basic understanding of the high resolution achieved in optical near-field microscopy as well as the large pseudomomentum exchange taking place in optical near-field interactions.

The electric field $\boldsymbol{E}^{(0)}$ from an oscillating electric dipole can be expanded in terms of a spherical basis as

$$\boldsymbol{E}^{(0)}(\boldsymbol{r}) = \sum_{m=-1}^{+1} (-1)^m E_{-m}^{(0)}(\boldsymbol{r})\hat{\boldsymbol{e}}_{+m} \,, \tag{55}$$

where the expansion coeffcients, $E_{-m}^{(0)}$, are defined by

$$E_{-m}^{(0)}(\boldsymbol{r}) = \hat{\boldsymbol{e}}_{-m} \cdot \boldsymbol{E}^{(0)}(\boldsymbol{r}) \,. \tag{56}$$

We consider that the electric dipole at $\boldsymbol{R}$ is oriented in the z direction as $\boldsymbol{d} = d\hat{\boldsymbol{e}}_0$ and the electric field of radiation is observed at a point $\boldsymbol{r} = \boldsymbol{R} + z_0\hat{\boldsymbol{e}}_0$ with z_0 the distance of the dipole and the observation point. For the dipole oriented in the z direction the nonzero coefficient of the angular-spectrum representation is given by

$$E_0^{(0)}(\boldsymbol{R} + z_0\hat{\boldsymbol{e}}_0) = \left(\frac{iK^3 d}{4\pi\epsilon_0}\right) \int_0^1 \mathrm{d}s_z \left(1 - s_z^2\right) \exp\left(i\rho_0 s_z\right)$$

$$+ \left(\frac{K^3 d}{4\pi\epsilon_0}\right) \int_0^{+\infty} \mathrm{d}\xi_z \left(1 + \xi_z^2\right) \exp\left(-\rho_0 \xi_z\right) \quad \text{for } 0 < z_0 \,, \tag{57}$$

where the integration with respect to $\mathrm{d}s_x$ and $\mathrm{d}s_y$ is converted to that to $\mathrm{d}s_z$ and $\mathrm{d}\beta$ under the relations; $s_x = \sqrt{1 - s_z^2}\cos\beta$ and $s_y = \sqrt{1 - s_z^2}\sin\beta$. We denote the normalized distance $\rho_0 = Kz_0$.

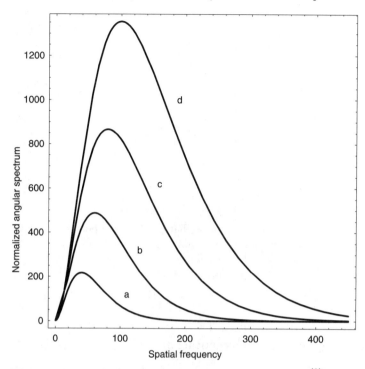

Fig. 8. The angular spectrum of evanescent waves for $E_0^{(0)}(\boldsymbol{R} + z_0\hat{\boldsymbol{e}}_0)$ normalized by the factor $K^3 d/(4\pi\epsilon_0)$ varsus the spatial frequency $\xi = -is_z$; a for $\rho_0 = 1/20$, b for $\rho_0 = 1/30$, c for $\rho_0 = 1/40$, d for $\rho_0 = 1/50$

When we consider electric dipoles described by $\boldsymbol{d} = d\hat{\boldsymbol{e}}_{\pm 1}$, the nonzero expansion coeffcients are those given by

$$E_{\mp 1}^{(0)}(\boldsymbol{R} + z_0\hat{\boldsymbol{e}}_0) = \left(\frac{iK^3 d}{4\pi\epsilon_0}\right) \int_0^1 ds_z \frac{-1}{2} \left(1 + s_z^2\right) \exp\left(i\rho_0 s_z\right)$$

$$+ \left(\frac{K^3 d}{4\pi\epsilon_0}\right) \int_0^{+\infty} d\xi_z \frac{1}{2} \left(\xi_z^2 - 1\right) \exp\left(-\rho_0 \xi_z\right) \quad \text{for } 0 < z_0 \, . \quad (58)$$

Each of the first terms in (57) and (58) represents the angular spectrum corresponding to homogeneous waves. The second terms are attributed to evanescent waves. Figure 8 shows the numerical example of the angular spectrum calculated for $E_0^{(0)}(\boldsymbol{R} + z_0\hat{\boldsymbol{e}}_0)$ normalized by the factor $K^3 d/(4\pi\epsilon_0)$. Since we consider the near-field regime $\rho_0 = Kz_0 \ll 1$, only the dominant parts of the angular spectra corresponding to evanescent waves are shown as the function of the normalized spatial frequency $\xi = -is_z$. It is clearly seen that the angular spectra take their maxima at the spatial frequencies corresponding to $\sim 1/\rho_0$ and have spectral width $\sim 2/\rho_0$. These results provide a very useful criterion to identify which spatial-frequency components are dominating the optical near-field process under consideration.

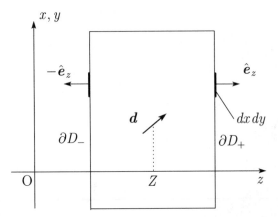

Fig. 9. Radiation power per second from an oscillating electric dipole is evaluated for a closed surface enclosing the source. By taking the height and depth of the enclosure in the x and y directions to be infinity, the total radiation power per second from the source is calculated as a surface integration of the Poynting vector on ∂D_+ and ∂D_-

3.4 Evaluation of Radiation Based on Angular-Spectrum Representation

In order to evaluate the radiation power per second from an oscillating electric dipole, we consider a closed surface ∂D enclosing the source. When there is no dissipation inside ∂D, the decrease in electromagnetic energy per second involved in ∂D is equivalent to the surface integration of the Poynting vector on the enclosure surface. We consider ∂D being composed of rectangular surfaces including xy-planes ∂D_+ and ∂D_-, as shown in Fig. 9. By taking the height and depth of the enclosure to be infinity for finite width, the total radiation power per second from the source is calculated as a surface integration of the Poynting vector on ∂D_+ and ∂D_-.

For the investigations of excitation transport between objects in the near-field regime, it is useful to evaluate separately the radiation power per second through ∂D_+ and ∂D_- as

$$I^{(0)} = I_+^{(0)} + I_-^{(0)} . \tag{59}$$

According to (27) and (28) the temporal averaged sum of the Poynting vector through each of the boundaries is obtained as

$$I_\pm^{(0)} = 2\epsilon_0 \int \int_{-\infty}^{+\infty} dx dy \Re \left\{ E^{(0)}(r) \times B^{(0)*}(r) \right\} \cdot (\pm \hat{e}_z) , \tag{60}$$

where $\pm \hat{e}_z$ stand for the unit normal vectors of the plane $\partial D_{\pm}$. Substituting (53) and (54) into (60), and evaluating the integration as

$$\int\int_{-\infty}^{+\infty} dx dy \exp\{iK[(s_x - s'_x)x + (s_y - s'_y)y]\}$$

$$= \left(\frac{2\pi}{K}\right)^2 \delta(s_x - s'_x)\delta(s_y - s'_y), \quad (61)$$

we obtain the total power per second through each boundary plane as

$$I_{\pm}^{(0)} = \left(\frac{K^4}{8\pi^2\epsilon_0}\right)^2 \sum_{\mu=1}^{2} \int\int_{0 \le s_{\parallel} < 1} \frac{ds_x ds_y}{s_z} \left[\boldsymbol{d}^* \cdot \hat{\boldsymbol{\varepsilon}}(\hat{\boldsymbol{s}}^{(\pm)}, \mu)\right]\left[\hat{\boldsymbol{\varepsilon}}(\hat{\boldsymbol{s}}^{(\pm)}, \mu) \cdot \boldsymbol{d}\right].$$

$$(62)$$

Here, we have utilized the following relations stand for the complex conjugate in accordance with (38);

$$s_z^* = \begin{cases} s_z & \text{for } 0 \le s_{\parallel} < 1, \\ -s_z & \text{for } 1 \le s_{\parallel} < +\infty, \end{cases} \quad (63)$$

or

$$\hat{\boldsymbol{s}}^{(\pm)*} = \begin{cases} \hat{\boldsymbol{s}}^{(\pm)} & \text{for } 0 \le s_{\parallel} < 1, \\ \hat{\boldsymbol{s}}^{(\mp)} & \text{for } 1 \le s_{\parallel} < +\infty. \end{cases} \quad (64)$$

It is noted that the radiation power through each of the boundary planes is due only to homogeneous waves, since evanescent waves can not propagate into the far-field region where existence of an optical sink is assumed in this problem. In other words, when we consider outgoing homogeneous waves we implicitly assume the existence of an optical sink, or reservoir in general, in the far-field region, otherwise any excited object is unable to radiate electromagnetic energy. This, in turn, implies that if an optical sink is put into the near-field region it will possibly alter the radiation properties of excited objects. This point is what we will discuss extensively later in this chapter and one of the most important issues in the study on functions and signal-transport properties of nanophotonic devices.

It is also useful to convert the integration with respect to Cartesian coordinates, ds_x and ds_y, into that described in terms of the spherical coordinates, α and β by utilizing the relations, $s_x = \sin\alpha\cos\beta$ and $s_y = \sin\alpha\sin\beta$;

$$I_{\pm}^{(0)} = \left(\frac{K^4}{8\pi^2\epsilon_0}\right)^2 \sum_{\mu=1}^{2} \int_0^{\pi/2} \sin\alpha d\alpha \int_0^{2\pi} d\beta \left[\boldsymbol{d}^* \cdot \hat{\boldsymbol{\varepsilon}}(\hat{\boldsymbol{s}}^{(\pm)}, \mu)\right]\left[\hat{\boldsymbol{\varepsilon}}(\hat{\boldsymbol{s}}^{(\pm)}, \mu) \cdot \boldsymbol{d}\right].$$

$$(65)$$

We then obtain the total power per second radiated from the oscillating electric dipole as

$$I^{(0)} = \left(\frac{K^4}{8\pi^2\epsilon_0}\right) \sum_{\mu=1}^{2} \int_0^{\pi} \sin\alpha\, d\alpha \int_0^{2\pi} d\beta \left[\boldsymbol{d}^* \cdot \hat{\boldsymbol{\varepsilon}}(\hat{\boldsymbol{s}}^{(+)}, \mu)\right] \left[\hat{\boldsymbol{\varepsilon}}(\hat{\boldsymbol{s}}^{(+)}, \mu) \cdot \boldsymbol{d}\right].$$
(66)

We can verify this result by making the scalar product of $\boldsymbol{d}^*$ with (52),

$$|\boldsymbol{d}|^2 = \left(\boldsymbol{d}^* \cdot \hat{\boldsymbol{s}}^{(\pm)}\right)\left(\hat{\boldsymbol{s}}^{(\pm)} \cdot \boldsymbol{d}\right) + \sum_{\mu=1}^{2}\left[\boldsymbol{d}^* \cdot \hat{\boldsymbol{\varepsilon}}(\hat{\boldsymbol{s}}^{(\pm)}, \mu)\right]\left[\hat{\boldsymbol{\varepsilon}}(\hat{\boldsymbol{s}}^{(\pm)}, \mu) \cdot \boldsymbol{d}\right], \quad (67)$$

and substituting this into the (+) component of (66), so that we obtain exactly the same result with (31). It is noted that in (66) $\hat{\boldsymbol{s}}^{(+)}$ corresponds to the direction of the observation of radiation in far field, i.e., $\hat{\boldsymbol{s}}^{(+)}$ can be replaced by $\hat{\boldsymbol{r}}_0$.

4   Radiative Decay of Oscillating Electric Dipole in Half-Space Based on Angular-Spectrum Representation

In this section we will study the classical theory of radiative decay of an oscillating electric dipole moment near a planar dielectric surface. We introduce a general treatment of half-space problems and discuss the fundamental processes involved in half-space problems. Then we will proceed to the theoretical evaluation of radiative decay of an oscillating electric dipole on the basis of angular-spectrum representation.

4.1   Half-Space Problems

Based on the results we have obtained using angular-spectrum representation of radiation fields in the previous section, we will study half-space problems with an actual planar boundary between two different dielectric media.

We consider a space, half of which is filled with a nonmagnetic, transparent, homogeneous, and isotropic dielectric medium of refractive index n (real number) and the other half is vacuum as shown in Fig. 10. The medium side is referred to as the left half-space, $z < 0$, and the vacuum side the right half-space, $z \geq 0$.

For later convenience, we will introduce unit wavevectors of incoming fields from the right of the boundary as $\hat{\boldsymbol{s}}^{(-)} = \boldsymbol{K}^{(-)}/K = (s_x, s_y, -s_z)$, and those of outgoing fields to the left of the boundary as $\hat{\boldsymbol{\kappa}}^{(-)} = \boldsymbol{k}^{(-)}/(nK) = (\kappa_x, \kappa_y, -\kappa_z)$, for which the following relations stand;

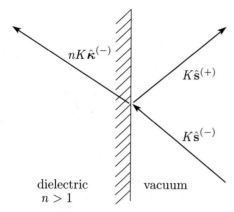

Fig. 10. Schematic diagram showing the half-space problems under consideration. The space, half of which is filled with a nonmagnetic, transparent, homogeneous, and isotropic dielectric medium of refractive index n (the left half-space, $z < 0$), and the other half is vacuum (the right half-space, $z \geq 0$)

$$\kappa_x = \frac{s_x}{n} , \tag{68}$$

$$\kappa_y = \frac{s_y}{n} , \tag{69}$$

$$\kappa_\| = \frac{s_\|}{n} , \tag{70}$$

$$\kappa_z = +\sqrt{1 - \kappa_\|^2} = +\frac{1}{n}\sqrt{n^2 - s_\|^2} , \tag{71}$$

$$s_z = +\sqrt{1 - s_\|^2} = +\sqrt{1 - n^2\kappa_\|^2} , \tag{72}$$

Here, the projections of the unit wavevectors onto the boundary surface, i.e., $\kappa_\| = \sqrt{\kappa_x^2 + \kappa_y^2}$, and $s_\| = \sqrt{s_x^2 + s_y^2}$, and $nK\kappa_\|$ and $Ks_\|$, correspond to the conserved quantities related to the translational symmetry of the system with respect to the planar boundary. In addition, we introduce unit wavevectors representing fields reflected from the boundary as $\hat{s}^{(+)} = \mathbf{K}^{(+)}/K = (s_x, s_y, s_z)$. The dispersion relations in terms of these variables are shown in Fig. 11.

According to (38), s_z is real for $0 \leq s_\| < 1$ and pure imaginary for $1 \leq s_\| < +\infty$, so that κ_z is represented by

$$\kappa_z = \begin{cases} +\dfrac{1}{n}\sqrt{n^2 - s_\|^2} & \text{for } 0 \leq s_\| < n , \\[4mm] +\dfrac{i}{n}\sqrt{s_\|^2 - n^2} & \text{for } n \leq s_\| < +\infty . \end{cases} \tag{73}$$

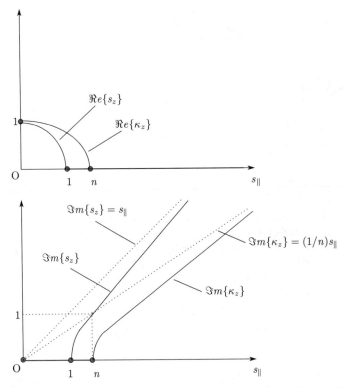

Fig. 11. Dispersion relations for unit wavevectors involved in half-space problems

Following the above classification, the light-scattering processes in the half-space configuration involve the following three characteristic cases, which are schematically shown in Fig. 12:

(i) When $0 \leq s_\| < 1$, $\hat{\boldsymbol{s}}^{(-)}$, $\hat{\boldsymbol{s}}^{(+)}$ and $\hat{\boldsymbol{\kappa}}^{(-)}$ are real vectors, therefore, the incident and reflected fields are homogeneous waves, the transmitted field is also a homogeneous wave. We introduce the incident angle α and transmitted angle α' in the domain,

$$0 \leq \alpha < \pi/2 \quad (s_\| = \sin \alpha),$$
$$0 \leq \alpha' < \alpha'_c \quad (\kappa_\| = \sin \alpha'),$$

where the angle α'_c corresponds to the critical angle of total internal reflection given by $\sin \alpha'_c = 1/n$. It is noted that the transmitted waves propagate in the direction within the critical angle of total internal reflection.

(ii) When $1 \leq s_\| < n$, $\hat{\boldsymbol{s}}^{(-)}$ and $\hat{\boldsymbol{s}}^{(+)}$ are complex vectors, and $\hat{\boldsymbol{\kappa}}^{(-)}$ is a real vector, therefore, the incident and reflected fields are evanescent waves, the transmitted field is a homogeneous wave. The incident and transmitted angles lie in the domain

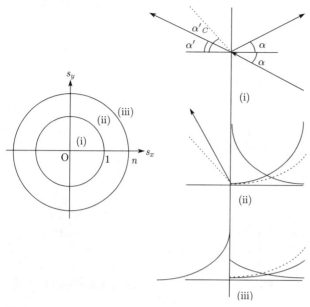

Fig. 12. The scattering process **(i)** for $0 \leq s_{\parallel} < 1$, corresponding to ordinary reflection and refraction of propagating wave. The scattering process **(ii)** for $1 \leq s_{\parallel} < n$, corresponding to radiation into a direction out of the angle of total internal reflection. The scattering process **(iii)** for $n \leq s_{\parallel}$, corresponding to optical near-field interactions localized near the boundary

$$\alpha = (\pi/2) - i\gamma , \quad 0 \leq \gamma < \gamma_c ,$$
$$\alpha'_c \leq \alpha' < \pi/2 .$$

Here, the angle γ_c corresponds to another critical angle set for extended fields given by $\sin[(\pi/2) - i\gamma_c] = n$. It is noted that the transmitted wave, in this case, propagates in the direction out of the critical angle of total internal reflection, α'_c, since $\alpha'_c \leq \alpha' < \pi/2$.

(iii) When $n \leq s_{\parallel} < +\infty$, $\hat{s}^{(-)}$, $\hat{s}^{(+)}$ and $\hat{\kappa}^{(-)}$ are complex vectors, therefore, the incident and reflected fields are evanescent waves, the transmitted field is also an evanescent wave. The incident and transmitted angles are in the domain of complex numbers,

$$\alpha = (\pi/2) - i\gamma , \quad \gamma_c \leq \gamma < +\infty ,$$
$$\alpha' = (\pi/2) - i\gamma' , \quad 0 \leq \gamma' < +\infty .$$

Therefore, the electromagnetic fields in this case are localized near the boundary. This case has no contribution to far-field observation, since no extended field is involved, however, it plays an important role in optical near-field interactions.

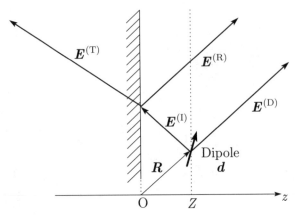

Fig. 13. The complex amplitude of electric field in half-space. $E^{(I)}$, $E^{(R)}$, and $E^{(T)}$ correspond to the incident, reflected and transmitted field, respectively. $E^{(D)}$ corresponds to the incident field into the right half-space of the dipole

4.2 Angular-Spectrum Representation of Radiation Fields in Half-Space

We consider a classical electric point dipole oscillating at frequency K located in the right half-space at R lying in the $z = 0$ plane as shown in Fig. 13. The radiation fields from the dipole into its left half-space, $z < Z$, are incident on the dielectric surface, for which the complex amplitude of the electric field is denoted as $E^{(I)}(r)$. According to (53), the angular-spectrum representation of $E^{(I)}(r)$ is given by

$$
E^{(I)}(r) = \left(\frac{iK^3}{8\pi^2\epsilon_0} \right) \sum_{\mu=1}^{2} \int\!\!\int_{-\infty}^{+\infty} ds_x ds_y \frac{1}{s_z}
$$
$$
\times \left[\hat{\varepsilon}(\hat{s}^{(-)}, \mu) \cdot d \right] \hat{\varepsilon}(\hat{s}^{(-)}, \mu) \exp\left[iK\hat{s}^{(-)} \cdot (r - R) \right] . \quad (74)
$$

The total complex amplitude of the electric field including contributions from the electric dipole and half-space system is obtained simply by summing up those of field components, $E^{(I)}(r)$, the radiation field from the dipole into its right half-space ($Z < z$), $E^{(D)}$, and the reflected field, $E^{(R)}$, and transmitted field, $E^{(T)}$ as

$$
E(r) = \begin{cases} E^{(R)}(r) + E^{(D)}(r) & \text{for } Z < z \,, \\[2mm] E^{(I)}(r) + E^{(R)}(r) & \text{for } 0 \le z < Z \,, \\[2mm] E^{(T)}(r) & \text{for } z < 0 \,. \end{cases} \quad (75)
$$

Using angular-spectrum representation of dipole fields, $E^{(I)}$, and Fresnel relations accounting for the electromagnetic boundary conditions, we can obtain the angular-spectrum representation of $E^{(R)}$ and $E^{(T)}$, respectively, by

$$\boldsymbol{E}^{(\mathrm{R})}(\boldsymbol{r}) = \left(\frac{\mathrm{i}K^3}{8\pi^2\epsilon_0}\right) \sum_{\mu=1}^{2} \int\!\!\int_{-\infty}^{+\infty} \mathrm{d}s_x \mathrm{d}s_y \frac{1}{s_z} \left[\hat{\boldsymbol{\varepsilon}}(\hat{\boldsymbol{s}}^{(-)}, \mu) \cdot \boldsymbol{d}\right]$$

$$\times \hat{\boldsymbol{\varepsilon}}(\hat{\boldsymbol{s}}^{(+)}, \mu) R_{\mathrm{R}}(s_z, \mu) \exp(\mathrm{i}K\hat{\boldsymbol{s}}^{(+)} \cdot \boldsymbol{r}) \exp(-\mathrm{i}K\hat{\boldsymbol{s}}^{(-)} \cdot \boldsymbol{R}) \ . \qquad (76)$$

$$\boldsymbol{E}^{(\mathrm{T})}(\boldsymbol{r}) = \left(\frac{\mathrm{i}K^3}{8\pi^2\epsilon_0}\right) \sum_{\mu=1}^{2} \int\!\!\int_{-\infty}^{+\infty} \mathrm{d}s_x \mathrm{d}s_y \frac{1}{s_z} \left[\hat{\boldsymbol{\varepsilon}}(\hat{\boldsymbol{s}}^{(-)}, \mu) \cdot \boldsymbol{d}\right]$$

$$\times \hat{\boldsymbol{\varepsilon}}(\hat{\boldsymbol{\kappa}}^{(-)}, \mu) T_{\mathrm{R}}(s_z, \mu) \exp(\mathrm{i}nK\hat{\boldsymbol{\kappa}}^{(-)} \cdot \boldsymbol{r}) \exp(-\mathrm{i}K\hat{\boldsymbol{s}}^{(-)} \cdot \boldsymbol{R}) \ . \qquad (77)$$

Here, R_{R} and T_{R} stand, respectively, for the reflection and transmission coefficients, for the incident wave from the right of the boundary given by

$$R_{\mathrm{R}}(s_z, \mu) = \begin{cases} \dfrac{s_z - n\kappa_z}{s_z + n\kappa_z} & \text{for } \mu = 1 \ , \\[2ex] \dfrac{ns_z - \kappa_z}{ns_z + \kappa_z} & \text{for } \mu = 2 \ , \end{cases} \qquad (78)$$

$$T_{\mathrm{R}}(s_z, \mu) = \begin{cases} \dfrac{2s_z}{s_z + n\kappa_z} & \text{for } \mu = 1 \ , \\[2ex] \dfrac{2s_z}{ns_z + \kappa_z} & \text{for } \mu = 2 \ , \end{cases} \qquad (79)$$

where $\kappa_z = (1/n)\sqrt{n^2 - 1 + s_z^2}$. In addition to the above, the electric field corresponding to direct radiation from the electric dipole into its right half-space, $z > Z$, is described as

$$\boldsymbol{E}^{(\mathrm{D})}(\boldsymbol{r}) = \left(\frac{\mathrm{i}K^3}{8\pi^2\epsilon_0}\right) \sum_{\mu=1}^{2} \int\!\!\int_{-\infty}^{+\infty} \mathrm{d}s_x \mathrm{d}s_y \frac{1}{s_z}$$

$$\times \left[\hat{\boldsymbol{\varepsilon}}(\hat{\boldsymbol{s}}^{(+)}, \mu) \cdot \boldsymbol{d}\right] \hat{\boldsymbol{\varepsilon}}(\hat{\boldsymbol{s}}^{(+)}, \mu) \exp\left[\mathrm{i}K\hat{\boldsymbol{s}}^{(+)} \cdot (\boldsymbol{r} - \boldsymbol{R})\right] \ . \qquad (80)$$

The complex amplitude for the total magnetic field associated with the electric field is obtained by the sum of field components as

$$\boldsymbol{B}(\boldsymbol{r}) = \begin{cases} \boldsymbol{B}^{(\mathrm{R})}(\boldsymbol{r}) + \boldsymbol{B}^{(\mathrm{D})}(\boldsymbol{r}) & \text{for } Z < z \ , \\[2ex] \boldsymbol{B}^{(\mathrm{I})}(\boldsymbol{r}) + \boldsymbol{B}^{(\mathrm{R})}(\boldsymbol{r}) & \text{for } 0 \leq z < Z \ , \\[2ex] \boldsymbol{B}^{(\mathrm{T})}(\boldsymbol{r}) & \text{for } z < 0 \ , \end{cases} \qquad (81)$$

where the complex field amplitudes are calculated from electric fields by using Maxwell's equation

$$\nabla \times \boldsymbol{E}(\boldsymbol{r}, t) = -\frac{\partial \boldsymbol{B}(\boldsymbol{r}, t)}{\partial t} \ . \qquad (82)$$

4.3 Electric Dipole Radiation into Medium

According to (75) and (81), the transmitted power per second of the dipole radiation into the medium, $I^{(-)}$, is given by

$$I_- = 2\epsilon_0 \int\int_{-\infty}^{+\infty} dx dy \Re\left\{\boldsymbol{E}^{(T)}(\boldsymbol{r}) \times \boldsymbol{B}^{(T)*}(\boldsymbol{r})\right\} \cdot (-\hat{\boldsymbol{e}}_z) . \qquad (83)$$

Substituting (77) and the associated magnetic field given by (82) and evaluating the following integration,

$$\int\int_{-\infty}^{+\infty} dx dy \exp\left\{inK[(\kappa_x - \kappa'_x)x + (\kappa_y - \kappa'_y)y]\right\}$$

$$= \left(\frac{2\pi}{K}\right)^2 \delta(s_x - s'_x)\delta(s_y - s'_y) , \qquad (84)$$

we obtain the transmitted power per second in the following form;

$$I_- = I_-^{(h)} + I_-^{(t)} , \qquad (85)$$

where $I_-^{(h)}$ corresponds to the integration for $0 \le s_\| < 1$ given by

$$I_-^{(h)} = \left(\frac{K^4}{8\pi^2\epsilon_0}\right) \sum_{\mu=1}^{2} \int\int_{0 \le s_\| < 1} \frac{ds_x ds_y}{s_z}$$

$$\times \left[\boldsymbol{d}^* \cdot \hat{\boldsymbol{\varepsilon}}(\hat{\boldsymbol{s}}^{(-)}, \mu)\right]\left[\hat{\boldsymbol{\varepsilon}}(\hat{\boldsymbol{s}}^{(-)}, \mu) \cdot \boldsymbol{d}\right] n\left(\frac{\kappa_z}{s_z}\right) T_R^2(s_z, \mu) \qquad (86)$$

and $I_-^{(t)}$ corresponds to the integration for $1 \le s_\| < n$ given by

$$I_-^{(t)} = \left(\frac{K^4}{8\pi^2\epsilon_0}\right) \sum_{\mu=1}^{2} \int\int_{1 \le s_\| < n} \frac{ds_x ds_y}{s_z} n\left(\frac{-\kappa_z}{s_z}\right)\left[\boldsymbol{d}^* \cdot \hat{\boldsymbol{\varepsilon}}(\hat{\boldsymbol{s}}^{(+)}, \mu)\right]$$

$$\times \left[\hat{\boldsymbol{\varepsilon}}(\hat{\boldsymbol{s}}^{(-)}, \mu) \cdot \boldsymbol{d}\right] T_R(s_z, \mu) T_R(-s_z, \mu) \exp(2iK s_z Z) . \qquad (87)$$

Here, we have utilized the following relations that stand for the complex conjugate in accordance with (73),

$$\kappa_z^* = \begin{cases} \kappa_z & \text{for } 0 \le s_\| < n , \\ -\kappa_z & \text{for } n \le s_\| < +\infty , \end{cases} \qquad (88)$$

$$\hat{\boldsymbol{\kappa}}^{(-)*} = \begin{cases} \hat{\boldsymbol{\kappa}}^{(-)} & \text{for } 0 \le s_\| < n , \\ \hat{\boldsymbol{\kappa}}^{(+)} & \text{for } n \le s_\| < +\infty , \end{cases} \qquad (89)$$

where $\hat{\boldsymbol{\kappa}}^{(+)} = (\kappa_x, \kappa_y, \kappa_z)$ and $\hat{\boldsymbol{\kappa}}^{(-)} = (\kappa_x, \kappa_y, -\kappa_z)$.

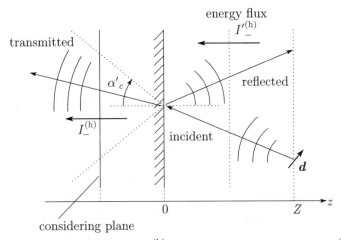

energy flux
$I'^{(\mathrm{h})}_-$

transmitted

α'_c

reflected

$I^{(\mathrm{h})}_-$

incident

d

0   Z z

considering plane

Fig. 14. The energy flux $I'^{(\mathrm{h})}_-$ corresponds to the energy flux $I^{(\mathrm{h})}_-$ into the medium in directions within the critical angle α'_{C}

According to the classification of fundamental processes shown in Fig. 12, $I^{(\mathrm{h})}_-$ corresponds to the case (i) since the unit wavevector $\hat{\boldsymbol{\kappa}}^{(-)}$ is a real vector for $0 \le \kappa_\parallel < 1/n$ ($0 \le s_\parallel < 1$). Therefore, $I^{(\mathrm{h})}_-$ corresponds to the energy flux transmitted into the medium in the directions lying within the critical angle of total internal reflection as shown in Fig. 14. On the other hand, $I^{(\mathrm{t})}_-$ corresponds to the case (ii) shown in Fig. 12, since the unit wavevector $\hat{\boldsymbol{\kappa}}^{(-)}$ is a real vector and $\hat{\boldsymbol{s}}^{(\pm)}$ is pure imaginary for $0 \le \kappa_\parallel < 1$ ($1 \le s_\parallel < n$). Therefore, $I^{(\mathrm{t})}_-$ corresponds to the energy flux into the medium in directions lying out of the critical angle of total internal reflection as shown in Fig. 14. In this case, the incident and reflected fields correspond to evanescent waves.

We will show later, in Sect. 4.5, that the energy flux from the oscillating electric dipole into the dielectric medium can be viewed as a tunneling process of excitation energy. The case (iii) shown in Fig. 12 has no direct contribution to the energy flux into the medium because the transmitted wave is localized near the dielectric boundary and contributes to higher-order interactions between the electric dipole and medium.

4.4 Electric Dipole Radiation into the Vacuum-Side Half-Space

According to (75) and (81), we can calculate the total radiation power per second $I^{(+)}$ from the oscillating electric dipole into the vacuum-side half-space through an arbitrary plane at $Z < z$ as

$$I_+ = I^{(\mathrm{h})}_+ + I^{(\mathrm{c})}_+ , \tag{90}$$

where $I^{(\mathrm{h})}_+$ corresponds to the direct sum of contributions from the radiated and reflected waves from the dipole given by

$$I_+^{(h)} = 2\epsilon_0 \int\int_{-\infty}^{+\infty} \mathrm{d}x\mathrm{d}y \Re e \left\{ \boldsymbol{E}^{(D)}(\boldsymbol{r}) \times \boldsymbol{B}^{(D)*}(\boldsymbol{r}) \right.$$

$$\left. + \boldsymbol{E}^{(R)}(\boldsymbol{r}) \times \boldsymbol{B}^{(R)*}(\boldsymbol{r}) \right\} \cdot \hat{\boldsymbol{e}}_z , \qquad (91)$$

and $I_+^{(c)}$ represents the modulation of radiation due to interference between direct and reflective waves given by

$$I_+^{(c)} = 2\epsilon_0 \int\int_{-\infty}^{+\infty} \mathrm{d}x\mathrm{d}y \Re e \left\{ \boldsymbol{E}^{(D)}(\boldsymbol{r}) \times \boldsymbol{B}^{(R)*}(\boldsymbol{r}) \right.$$

$$\left. + \boldsymbol{E}^{(R)}(\boldsymbol{r}) \times \boldsymbol{B}^{(D)*}(\boldsymbol{r}) \right\} \cdot \hat{\boldsymbol{e}}_z . \qquad (92)$$

Using (76), (80)–(82), and (61), we obtain

$$I_+^{(h)} = \left(\frac{K^4}{8\pi^2\epsilon_0} \right) \sum_{\mu=1}^{2} \int\int_{0 \leq s_\| < 1} \mathrm{d}s_x \mathrm{d}s_y \frac{1}{s_z}$$

$$\times \left\{ \left[\boldsymbol{d}^* \cdot \hat{\boldsymbol{\varepsilon}}(\hat{\boldsymbol{s}}^{(+)}, \mu) \right] \left[\hat{\boldsymbol{\varepsilon}}(\hat{\boldsymbol{s}}^{(+)}, \mu) \cdot \boldsymbol{d} \right] \right.$$

$$\left. + \left[\boldsymbol{d}^* \cdot \hat{\boldsymbol{\varepsilon}}(\hat{\boldsymbol{s}}^{(-)}, \mu) \right] \left[\hat{\boldsymbol{\varepsilon}}(\hat{\boldsymbol{s}}^{(-)}, \mu) \cdot \boldsymbol{d} \right] R_R^2(s_z, \mu) \right\} , \qquad (93)$$

$I_+^{(c)}$ can be obtained as

$$I_+^{(c)} = \left(\frac{K^4}{4\pi^2\epsilon_0} \right) \Re e \left\{ \sum_{\mu=1}^{2} \int\int_{0 \leq s_\| < 1} \mathrm{d}s_x \mathrm{d}s_y \frac{1}{s_z} \right.$$

$$\left. \times \left[\boldsymbol{d}^* \cdot \hat{\boldsymbol{\varepsilon}}(\hat{\boldsymbol{s}}^{(+)}, \mu) \right] \left[\hat{\boldsymbol{\varepsilon}}(\hat{\boldsymbol{s}}^{(-)}, \mu) \cdot \boldsymbol{d} \right] R_R(s_z, \mu) \exp(2iKs_zZ) \right\} . \qquad (94)$$

The interference term involves contributions only from homogeneous waves. Since the factor $\exp(2iKs_zZ)$ for $0 \leq s_\| < 1$, oscillates rapidly when the distance between the dipole and the dielectric surface becomes large compared with the optical wavelength, the interference term disappears, $I_+^{(c)} \to 0$, for $1 \ll KZ$.

4.5   Interaction between Electric Dipole and Dielectric Surface

We consider optical near-field interactions of the oscillating electric dipole with the planar dielectric surface in terms of the energy flux through a plane assumed to be in between the electric dipole and the surface, $0 < z < Z$. It will be shown that optical near-field interactions are described in terms of the Poynting vector composed of cross-products of evanescent waves, one being radiation from the dipole and another from the surface. One can find such a cross-product of evanescent waves in the description of tunneling current where electronic scalar wave functions are penetrating into a barrier region with exponential decay. This implies that we can picture an optical near-field interaction as a tunneling process of an optical excitation. The result derived in the following provides one of the most important bases in the study of optical near-field problems.

According to (75) and (81), we obtain the total power per second, I'_-, through the assumed plane from the dipole to the surface as

$$I'_- = I'^{(h)}_- + I'^{(t)}_- , \tag{95}$$

where $I'^{(h)}_-$ corresponds to contributions from both the radiated and reflective fields given by

$$I'^{(h)}_- = 2\epsilon_0 \int\!\!\int_{-\infty}^{+\infty} \mathrm{d}x\mathrm{d}y \Re e \left\{ \boldsymbol{E}^{(\mathrm{I})}(\boldsymbol{r}) \times \boldsymbol{B}^{(\mathrm{I})*}(\boldsymbol{r}) \right.$$
$$\left. + \boldsymbol{E}^{(\mathrm{R})}(\boldsymbol{r}) \times \boldsymbol{B}^{(\mathrm{R})*}(\boldsymbol{r}) \right\} \cdot (-\hat{\boldsymbol{e}}_z) , \tag{96}$$

and $I'^{(t)}_-$ describes the optical near-field interaction of the electric dipole with the dielectric surface given by

$$I'^{(t)}_- = 2\epsilon_0 \int\!\!\int_{-\infty}^{+\infty} \mathrm{d}x\mathrm{d}y \Re e \left\{ \boldsymbol{E}^{(\mathrm{I})}(\boldsymbol{r}) \times \boldsymbol{B}^{(\mathrm{R})*}(\boldsymbol{r}) \right.$$
$$\left. + \boldsymbol{E}^{(\mathrm{R})}(\boldsymbol{r}) \times \boldsymbol{B}^{(\mathrm{I})*}(\boldsymbol{r}) \right\} \cdot (-\hat{\boldsymbol{e}}_z) . \tag{97}$$

Using (74), (76), (81), (82) and (61), we obtain $I'^{(h)}_-$ as

$$I'^{(h)}_- = \left(\frac{K^4}{8\pi^2\epsilon_0} \right) \sum_{\mu=1}^{2} \int\!\!\int_{0 \leq s_\parallel < 1} \mathrm{d}s_x \mathrm{d}s_y \frac{1}{s_z}$$
$$\times \left[\boldsymbol{d}^* \cdot \hat{\boldsymbol{\varepsilon}}(\hat{\boldsymbol{s}}^{(-)}, \mu) \right] \left[\hat{\boldsymbol{\varepsilon}}(\hat{\boldsymbol{s}}^{(-)}, \mu) \cdot \boldsymbol{d} \right] \left[1 - R_\mathrm{R}^2(s_z, \mu) \right] , \tag{98}$$

which simply describe the net energy transport via homogeneous waves propagating from right to left and left to right. It is noted that conservation of energy flux holds as

$$I'^{(h)}_- = I^{(h)}_- , \tag{99}$$

which is verified by using the relation between reflection and transmission coefficients given as

$$T_\mathrm{R}^2(s_z, \mu) = \left(\frac{s_z}{n\kappa_z} \right) \left[1 - R_\mathrm{R}^2(s_z, \mu) \right] . \tag{100}$$

As we discussed in Fig. 14, the net energy flux flow into a medium due to homogeneous waves corresponds to that carried by homogeneous waves propagating in the direction within the critical angle of total internal reflection.

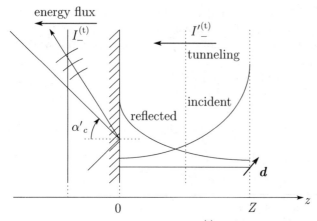

energy flux

Fig. 15. The tunneling energy flux $I'^{(t)}_-$ corresponds to the energy flux $I^{(t)}_-$ into the medium in directions without the critical angle α'_C

In the same manner as the above, we obtain $I'^{(t)}_-$ as

$$
I'^{(t)}_- = \left(\frac{K^4}{4\pi^2\epsilon_0}\right)\Re e\left\{\sum_{\mu=1}^{2}\int\int_{1\le s_\parallel <n}\mathrm{d}s_x\mathrm{d}s_y\frac{1}{s_z}\right.
$$
$$
\left. \times\left[\boldsymbol{d}^*\cdot\hat{\boldsymbol{\varepsilon}}(\hat{\boldsymbol{s}}^{(+)},\mu)\right]\left[\hat{\boldsymbol{\varepsilon}}(\hat{\boldsymbol{s}}^{(-)},\mu)\cdot\boldsymbol{d}\right]R_{\mathrm{R}}(s_z,\mu)\exp(2\mathrm{i}Ks_zZ)\right\}, \quad (101)
$$

which is due to near-field electromagnetic interactions between the planar dielectric boundary and electric dipole via evanescent waves with $1\le s_\parallel <n$, so that $I'^{(t)}_-$ can be interpreted as the tunnel current for transport of optical energy. Since the factor $\exp(2\mathrm{i}Ks_zZ) = \exp(-2K|s_z|Z)$ for $1\le s_\parallel <n$ exhibits an exponential decay with increase in the dipole-to-surface distance, the tunnel effect is important only in the near-field regime. Conservation of energy flux,

$$
I'^{(t)}_- = I^{(t)}_- , \quad (102)
$$

is verified by using the relation between the reflection and transmission coefficients;

$$
T_{\mathrm{R}}(-s_z,\mu)T_{\mathrm{R}}(s_z,\mu) = \left(\frac{s_z}{n\kappa_z}\right)[R_{\mathrm{R}}(-s_z,\mu) - R_{\mathrm{R}}(s_z,\mu)] . \quad (103)
$$

As we discussed in Fig. 15, the energy transport due to tunneling of excitation results in homogeneous waves in the medium with propagation direction falling in the angular domain out of the critical angle of total internal reflection.

It should be noted that (93) and (98) leads to the relation

$$
I^{(h)}_+ + I'^{(h)}_- = I^{(0)} . \quad (104)
$$

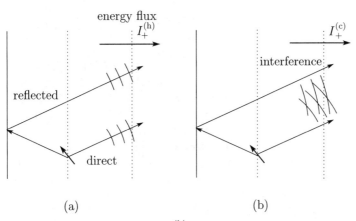

Fig. 16. (a) The energy flux $I_+^{(h)}$ composed of contributions from direct, $\boldsymbol{E}^{(D)}$, and reflected, $\boldsymbol{E}^{(R)}$, fields. (b) The energy flux $I_+^{(c)}$ corresponding to the intensity modulation due to interference between direct and reflected fields

This implies an important fact that the radiation power per second from an oscillating electric dipole in free space, $I^{(0)}$, is equal to that from the same dipole near a planar dielectric surface via homogeneous waves. As we have seen in the half-space problem, there remains an excess radiation energy, $I'^{(t)}_-$ and $I_+^{(c)}$, from the dipole near the surface due, respectively, to tunneling of optical excitation via evanescent waves and interference between the direct and reflected homogeneous waves. The excess radiation power per second of the dipole,

$$\Delta I = I_+^{(c)} + I'^{(t)}_- , \tag{105}$$

suggests that the radiative decay rate of the oscillating electric dipole is modified due to optical near-field interaction with the dielectric surface. This important feature of optical near-field interactions is fully developed later in this chapter on the basis of 2nd-quantization of optical fields in half-space problems.

5 Quantum Theory of Dipole Radiation Near a Dielectric Surface Based on Detector Modes

In general, quantum-mechanical treatment of fields is achieved in the second quantization framework based on normal modes. Normal modes should be composed so as to describe the state of the fields in the entire space considered in the problem. Since half-space problems involve two different regions, whether these regions are different or not in optical properties, we should compose the normal modes so as to include fields of both sides of the half-space under some adequate connections of the fields at the boundary plane.

For the basis of field quantization including evanescent waves, Carniglia and Mandel [22] introduced the so-called triplet mode composed of a set of incident, reflected, and transmitted waves connected via Fresnel's relations at the planar boundary under consideration. The triplet modes involving a single incident wave serve as the convenient basis for the theoretical treatment of photon-absorption processes near a dielectric surface, where a single light source placed in far field may be assumed in a practical setup. In contrast, in accounting for the photon-emission processes near the boundary, the triplet mode should be related to a correlation measurement of photons by using two photodetectors coupled to each of the outgoing waves involved in the triplet. This situation, in some sense, is similar to beam-splitter problems.

On the other hand, when we study photon-emission processes near a dielectric surface, we usually consider a practical setup with a photon-counting scheme by using an independent photodetector placed in each of the half spaces separated by the boundary, so that we may consider a single outgoing wave as the final state of the radiation process. In this case, the so-called detector modes including a single outgoing wave serve as the convenient basis for theoretical analysis, especially in the consideration of radiative lifetime of a two-level system near a dielectric surface. Indeed, completeness of basis functions assures equivalence between various descriptions, but practical treatments of photon absorption or emission experiments are simplified by choosing either of those expressions according to the experimental setup under consideration: either a single source or a single detector is assumed in the far-field region. The detector modes have been introduced by Viogureux and Payen [3] in their theoretical study on the Raman diffusion due to atoms near a planar dielectric boundary. In their work, the detector modes are defined in terms of the linear combination of the triplet mode functions but have not been explicitly quantized.

In this section, we study the field quantization based on the detector mode as the basis for theoretical analysis of the photon-emission process in the optical near field. In contrast to the study reported by Viogureux and Payen we introduce the detector modes in terms of time-reversal and spatial-rotation transforms of the triplet mode. This gives us a straightforward evaluation of the radiative decay rate in terms of the final-state density of the photonic mode and provides a clear understanding of the meaning of detector mode as well as the correspondence between classical and quantum descriptions of electric and magnetic multipoles radiations of a two-level system in the near-field regime.

5.1 Normal Modes as the Basis of Field Quantization in Half-Space Problems; Triplet and Detector Modes

In this section we consider a space, half of which is filled with a nonmagnetic, transparent, homogeneous, and isotropic dielectric medium of refractive index n (left half-space, $z < 0$), and half is vacuum (right half-space, $z \geq 0$) (see

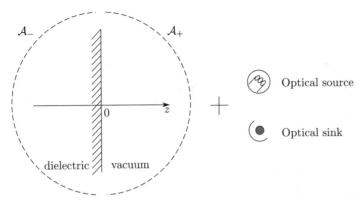

Fig. 17. Photonic system of half-space. Optical sources and sinks are considered to be placed on the left $\mathcal{A}_-$ and the right $\mathcal{A}_+$ hemispheres of radius $r_\pm$ in far field, $kr_\pm$, $Kr_\pm \gg 1$

Fig. 17). Optical sources and sinks are considered to be placed on the left $\mathcal{A}_-$ and the right $\mathcal{A}_+$ hemispheres of radius $r_\pm$ in far field, $kr_\pm$, $Kr_\pm \gg 1$. In this limit, we consider the space inside the whole sphere, $\mathcal{A}_+$ plus $\mathcal{A}_-$, as the entire space for which normal modes are defined. Corresponding to these, when we introduce optical sources and sinks in far field on $\mathcal{A}_-$ and $\mathcal{A}_+$ hemispheres, they are considered, respectively, as emitters and detectors of incoming and outgoing plane waves.

We will study the problem for the monochromatic fields of frequency K corresponding to the Fourier component of electric field $\boldsymbol{E}(\boldsymbol{r})\exp(-iKt)$, under the unit in which the light velocity is taken to be unity, $c = 1$. The complex amplitude $\boldsymbol{E}(\boldsymbol{r})$ of the electric field satisfies the Helmholtz equation

$$\left[\nabla^2 + K^2 n^2(\boldsymbol{r})\right]\boldsymbol{E}(\boldsymbol{r}) = \boldsymbol{0} \,, \tag{106}$$

with the refractive-index function defined by

$$n(\boldsymbol{r}) = \begin{cases} n & \text{for } z < 0 \,, \\ 1 & \text{for } z \geq 0 \,, \end{cases} \tag{107}$$

where n is assumed to be a real number.

As the basis to introduce normal modes in half-space problems, we introduce the unit wavevectors of incoming waves from the right of the boundary plane as $\hat{\boldsymbol{s}}^{(-)} = \boldsymbol{K}^{(-)}/K = (s_x, s_y, -s_z)$, and those of outgoing fields to the left of the boundary plane as $\hat{\boldsymbol{\kappa}}^{(-)} = \boldsymbol{k}^{(-)}/(nK) = (\kappa_x, \kappa_y, -\kappa_z)$, which satisfy the relations in (68)–(72). We also introduce the unit wavevectors of incoming waves from the left as $\hat{\boldsymbol{\kappa}}^{(+)} = \boldsymbol{k}^{(+)}/nK = (\kappa_x, \kappa_y, \kappa_z)$ and those of the outgoing field to the right as $\hat{\boldsymbol{s}}^{(+)} = \boldsymbol{K}^{(+)}/K = (s_x, s_y, s_z)$. The projection of the wavevector onto the boundary plane is conserved due to spatial

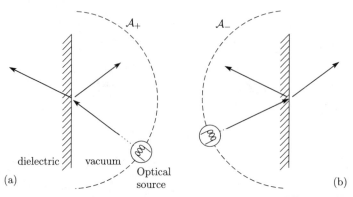

Fig. 18. The triplet modes. **(a)** R-triplet mode coupled to a single optical source on the hemisphere $\mathcal{A}_+$. **(b)** L-triplet mode coupled to a single optical source on the hemisphere $\mathcal{A}_-$

translational symmetry in half-space problems, so that we introduce $nK\kappa_\parallel$ and $Ks_\parallel$ defined, respectively, by $\kappa_\parallel = \sqrt{\kappa_x^2 + \kappa_y^2}$, and $s_\parallel = \sqrt{s_x^2 + s_y^2}$.

When we consider a problem with a single optical source, it is convenient to employ the so-called triplet modes composed of one incoming and two outgoing plane waves that are connected at the boundary plane by using Fresnel's relations for fulfillment of electromagnetic boundary conditions as shown in Figs. 18a–b.

For instance, when we consider absorption properties of atoms put in the vacuum side of the half-space, we consider that the incident wave of the triplet mode is coupled to a light source placed at far field and evaluate the electric-dipole interaction of atoms with the wave components in the vacuum side involved in the triplet mode. The triplet modes are also applicable to a problem in which a photonic source, such as an excited atom, is put in the half-space configuration under consideration. For instance, when we put an excited atom in a subwavelength vicinity of the dielectric surface, we can evaluate the radiation properties of the atom in terms of interactions with the triplet mode. It should, however, be noted that, in the triplet-mode configuration, the resulting two outgoing waves corresponding to a single-photon emission are considered to have quantum correlation. Such a property is similar to those we encounter as beam-splitter problems in quantum optics. Therefore, the triplet-mode description has convenience in the theoretical treatments of a half-space photonic system in terms of a correlation measurement of radiation by a pair of photodetectors put, respectively, on the hemispheres $\mathcal{A}_-$ and $\mathcal{A}_+$.

As the complementary description of the half-space problems, we can introduce so-called detector modes as the normal modes composed of a single outgoing and two incident plane waves connected by Fresnel's relations at the boundary surface. The detector-mode description is especially useful in investigations on radiation properties of photonic source, such as an excited atom, put in the half-space system, since a single outgoing wave coupled to a detector in far field is involved (see Figs. 19a–b). For example, when we investigate the spontaneous emission of radiation from an excited atom put near a dielectric surface using a single photodetector put on one of the hemispheres, $\mathcal{A}_-$ or $\mathcal{A}_+$, we can evaluate the radiation properties in terms of the atomic interaction with the wave components in the vacuum side. In this case, we can easily evaluate the spontaneous emission by virtue of the well-defined final-state mode-density of radiation involved in the corresponding one of the two detector modes.

In the following, we introduce the detector-mode functions by means of time reversal and spatial rotation transform of the well-established triplet-mode functions, for which orthogonality relations and completeness are confirmed by the work of Carniglia and Mandel [22]. We will see that the detector-mode description of radiation problems in half-space provides us with a clear interpretation of the fundamental processes that can be compared directly with classical descriptions of radiation, such as image-dipole pictures useful in understanding half-space problems. It is noted that the time-reversal and spatial-rotation transform employed preserves the field momentum parallel to the boundary surface and, therefore, the angular momentum of the field normal to the surface, which are the conserved quantities in the interacting atom plus photon system under consideration. In the sense of restricted conservation laws, we refer to these quantities as pseudomomentum and angular pseudomomentum, respectively. These conserved quantities under the restricted symmetry of the local system must play the most important role in optical manipulation and control of the mesoscopic material system of nanometer size as well as in realization of functional devices working on the basis of optoelectronic interactions.

5.2   Detector-Mode Functions

As is shown in Fig. 19a, the R-detector-mode function is defined in terms of a composition of three field amplitudes by

$$\mathcal{E}_{\mathrm{DR}}(\boldsymbol{K}^{(+)}, \mu, \boldsymbol{r}) = \mathcal{E}_{\mathrm{DR}}^{(\mathrm{I})}(\boldsymbol{K}^{(+)}, \mu, \boldsymbol{r}) + \mathcal{E}_{\mathrm{DR}}^{(\mathrm{R})}(\boldsymbol{K}^{(-)}, \mu, \boldsymbol{r})$$
$$+ \mathcal{E}_{\mathrm{DR}}^{(\mathrm{T})}(\boldsymbol{k}^{(+)}, \mu, \boldsymbol{r}) \,, \quad (108)$$

where DR indicates that the mode involves a single outgoing wave in vacuum, i.e., right-half to the boundary, and the parameters $\boldsymbol{K}^{(+)}$ and μ in the field amplitudes $\mathcal{E}_{\mathrm{DR}}$ indicate, respectively, the wavevector and the polarization state of the outgoing wave into vacuum; $\mu = 1$ for the transverse electric (TE) polarization and $\mu = 2$ for the transverse magnetic (TM) polarization. Using the notation of vector plane waves, we can describe the field components, i.e., single outgoing and two incident waves, as follows;

$$\mathcal{E}_{\mathrm{DR}}^{(\mathrm{I})}(\boldsymbol{K}^{(+)}, \mu, \boldsymbol{r})$$
$$= \begin{cases} \dfrac{1}{\sqrt{2}} \hat{\varepsilon}(\hat{\boldsymbol{s}}^{(+)}, \mu) \exp(\mathrm{i}K\hat{\boldsymbol{s}}^{(+)} \cdot \boldsymbol{r}) & \text{for } z \geq 0 \,, \\[2mm] 0 & \text{for } z < 0 \,, \end{cases} \tag{109}$$

$$\mathcal{E}_{\mathrm{DR}}^{(\mathrm{R})}(\boldsymbol{K}^{(-)}, \mu, \boldsymbol{r})$$
$$= \begin{cases} \dfrac{1}{\sqrt{2}} \hat{\varepsilon}(\hat{\boldsymbol{s}}^{(-)}, \mu) R_{\mathrm{R}}(s_z, \mu) \exp(\mathrm{i}K\hat{\boldsymbol{s}}^{(-)} \cdot \boldsymbol{r}) & \text{for } z \geq 0 \,, \\[2mm] 0 & \text{for } z < 0 \,, \end{cases}$$
$$\tag{110}$$

$$\mathcal{E}_{\mathrm{DR}}^{(\mathrm{T})}(\boldsymbol{k}^{(+)}, \mu, \boldsymbol{r})$$
$$= \begin{cases} \dfrac{1}{\sqrt{2}} \hat{\varepsilon}(\hat{\boldsymbol{\kappa}}^{(+)}, \mu) T_{\mathrm{R}}(s_z, \mu) \exp(\mathrm{i}nK\hat{\boldsymbol{\kappa}}^{(+)} \cdot \boldsymbol{r}) & \text{for } z < 0 \,, \\[2mm] 0 & \text{for } z \geq 0 \,. \end{cases}$$
$$\tag{111}$$

Here, $R_{\mathrm{R}}(s_z, \mu)$ and $T_{\mathrm{R}}(s_z, \mu)$ are, respectively, the reflection and transmission coefficients of incident waves from right to left given by (78) and (79). Since the wave component $\mathcal{E}_{\mathrm{DR}}^{(\mathrm{I})}$ is considered to be coupled with a single photodetector placed on the right hemisphere $\mathcal{A}_+$ in far field, the unit wavevector $\hat{\boldsymbol{s}}^{(+)}$ falls in the domain where $0 \leq s_\| < 1$ ($0 \leq \kappa_\| < 1/n$), so that $\hat{\boldsymbol{s}}^{(+)}$, $\hat{\boldsymbol{s}}^{(-)}$ and $\hat{\boldsymbol{\kappa}}^{(+)}$ are always real vectors and $\mathcal{E}_{\mathrm{DR}}$ corresponds to homogeneous waves.

According to the diagram shown in Fig. 19b, the L-detector-mode function is defined by

$$\mathcal{E}_{\mathrm{DL}}(\boldsymbol{k}^{(-)}, \mu, \boldsymbol{r}) = \mathcal{E}_{\mathrm{DL}}^{(\mathrm{I})}(\boldsymbol{k}^{(-)}, \mu, \boldsymbol{r}) + \mathcal{E}_{\mathrm{DL}}^{(\mathrm{R})}(\boldsymbol{k}^{(+)}, \mu, \boldsymbol{r})$$
$$+ \mathcal{E}_{\mathrm{DL}}^{(\mathrm{T})}(\boldsymbol{K}^{(-)*}, \mu, \boldsymbol{r}) \,, \tag{112}$$

where DL indicates that the mode involves a single outgoing wave in the medium, i.e., the left-half to the boundary, and the parameters $\boldsymbol{k}^{(-)}$ and μ in $\mathcal{E}_{\mathrm{DL}}$ indicate, respectively, the wavevector and polarization state of the outgoing wave in medium. We can describe the field components as

$$\mathcal{E}_{\mathrm{DL}}^{(\mathrm{I})}(\boldsymbol{k}^{(-)}, \mu, \boldsymbol{r})$$
$$= \begin{cases} \dfrac{1}{\sqrt{2}n} \hat{\varepsilon}(\hat{\boldsymbol{\kappa}}^{(-)}, \mu) \exp(\mathrm{i}nK\hat{\boldsymbol{\kappa}}^{(-)} \cdot \boldsymbol{r}) & \text{for } z < 0 \,, \\[2mm] 0 & \text{for } z \geq 0 \,, \end{cases}$$
$$\tag{113}$$

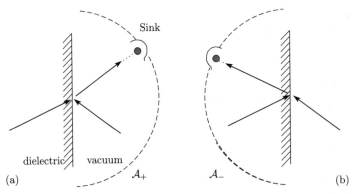

Fig. 19. The detector-mode functions in half-space problems. (**a**) R detector-mode function coupled to a single optical sink placed on the hemisphere $\mathcal{A}_+$ in far field. (**b**) L detector-mode function coupled to that on $\mathcal{A}_-$

$$
\mathcal{E}_{\mathrm{DL}}^{(\mathrm{R})}(\boldsymbol{k}^{(+)}, \mu, \boldsymbol{r})
$$
$$
= \begin{cases} \dfrac{1}{\sqrt{2}n}\hat{\boldsymbol{\varepsilon}}(\hat{\boldsymbol{\kappa}}^{(+)}, \mu)R_{\mathrm{L}}(s_z^*, \mu)\exp(\mathrm{i}nK\hat{\boldsymbol{\kappa}}^{(+)}\cdot\boldsymbol{r}) & \text{for } z < 0 \,, \\[2ex] \boldsymbol{0} & \text{for } z \geq 0 \,, \end{cases}
$$
$$
\tag{114}
$$

$$
\mathcal{E}_{\mathrm{DL}}^{(\mathrm{T})}(\boldsymbol{K}^{(-)*}, \mu, \boldsymbol{r})
$$
$$
= \begin{cases} \dfrac{1}{\sqrt{2}n}\hat{\boldsymbol{\varepsilon}}(\hat{\boldsymbol{s}}^{(-)*}, \mu)T_{\mathrm{L}}(s_z^*, \mu)\exp(\mathrm{i}K\hat{\boldsymbol{s}}^{(-)*}\cdot\boldsymbol{r}) & \text{for } z \geq 0 \,, \\[2ex] \boldsymbol{0} & \text{for } z < 0 \,. \end{cases}
$$
$$
\tag{115}
$$

Here, $R_{\mathrm{L}}(s_z, \mu)$ and $T_{\mathrm{L}}(s_z, \mu)$ are, respectively, the reflection and transmission coefficients for the incident wave from the medium side given by

$$
R_{\mathrm{L}}(s_z, \mu) = \begin{cases} \dfrac{n\kappa_z - s_z}{n\kappa_z + s_z} & \text{for } \mu = 1 \,, \\[2ex] \dfrac{\kappa_z - ns_z}{\kappa_z + ns_z} & \text{for } \mu = 2 \,, \end{cases}
\tag{116}
$$

$$
T_{\mathrm{L}}(s_z, \mu) = \begin{cases} \dfrac{2n\kappa_z}{n\kappa_z + s_z} & \text{for } \mu = 1 \,, \\[2ex] \dfrac{2n\kappa_z}{\kappa_z + ns_z} & \text{for } \mu = 2 \,. \end{cases}
\tag{117}
$$

Since the wave component $\mathcal{E}_{\mathrm{DL}}^{(\mathrm{I})}$ is to be coupled with a single photodetector on $\mathcal{A}_-$, the unit wavevector $\hat{\boldsymbol{\kappa}}^{(-)}$ falls in the domain of $0 \leq \kappa_\| < 1$ ($0 \leq$

$s_\parallel < n$), so that $\hat{\boldsymbol{\kappa}}^{(-)}$ and $\hat{\boldsymbol{\kappa}}^{(+)}$ are real vectors, and $\hat{\boldsymbol{s}}^{(-)*}$ involves a complex component. For $0 \leq \kappa_\parallel < 1/n$ ($0 \leq s_\parallel < 1$), $-s_z$ is real, so that the relations

$$\hat{\boldsymbol{s}}^{(-)*} = \hat{\boldsymbol{s}}^{(-)} , \quad -s_z^* = -s_z ,$$

hold and $\mathcal{E}_{\mathrm{DL}}$ corresponds to homogeneous waves. On the other hand, for $1/n \leq \kappa_\parallel < 1$ ($1 \leq s_\parallel < n$), $-s_z$ is purely imaginary, so that

$$\hat{\boldsymbol{s}}^{(-)*} = \hat{\boldsymbol{s}}^{(+)} , \quad -s_z^* = s_z ,$$

and $\mathcal{E}_{\mathrm{DL}}$ corresponds to the evanescent wave.

In a practical setup, $\boldsymbol{k}^{(-)}$ and $\boldsymbol{K}^{(+)}$, indicating the outgoing wavevectors of detector modes, correspond to the angular directions of the light sink placed on $\mathcal{A}_-$ and $\mathcal{A}_+$, respectively. L and R detector modes correspond to the eigenstates of the pseudomomentum operator $-i\hbar\nabla_\parallel$ as

$$(-i\hbar\nabla_\parallel)\mathcal{E}_{\mathrm{DR}}(\boldsymbol{K}^{(+)}, \mu, \boldsymbol{r}) = (\hbar K \hat{\boldsymbol{s}}_\parallel)\mathcal{E}_{\mathrm{DR}}(\boldsymbol{K}^{(+)}, \mu, \boldsymbol{r}) , \tag{118}$$

$$(-i\hbar\nabla_\parallel)\mathcal{E}_{\mathrm{DL}}(\boldsymbol{k}^{(-)}, \mu, \boldsymbol{r}) = (\hbar K \hat{\boldsymbol{s}}_\parallel)\mathcal{E}_{\mathrm{DL}}(\boldsymbol{k}^{(-)}, \mu, \boldsymbol{r}) , \tag{119}$$

where $-i\hbar\nabla_\parallel$ operates on each component of $\mathcal{E}$. We have denoted the pseudomomentum vector $\hbar K \hat{\boldsymbol{s}}_\parallel = \hbar K(s_x, s_y, 0)$.

The orthogonality relations for the detector-mode functions are obtained as follows;

$$\int\int\int \mathrm{d}^3x n^2(\boldsymbol{r}) \left[\mathcal{E}_{\mathrm{DR}}(\boldsymbol{K}^{(+)}, \mu, \boldsymbol{r})\right]^* \cdot \mathcal{E}_{\mathrm{DR}}(\boldsymbol{K}'^{(+)}, \mu', \boldsymbol{r})$$

$$= \frac{1}{2}(2\pi)^3 \delta_{\mu\mu'}\delta^3(\boldsymbol{K}^{(+)} - \boldsymbol{K}'^{(+)}) , \tag{120}$$

$$\int\int\int \mathrm{d}^3x n^2(\boldsymbol{r}) \left[\mathcal{E}_{\mathrm{DL}}(\boldsymbol{k}^{(-)}, \mu, \boldsymbol{r})\right]^* \cdot \mathcal{E}_{\mathrm{DL}}(\boldsymbol{k}'^{(-)}, \mu', \boldsymbol{r})$$

$$= \frac{1}{2}(2\pi)^3 \delta_{\mu\mu'}\delta^3(\boldsymbol{k}^{(-)} - \boldsymbol{k}'^{(-)}) , \tag{121}$$

$$\int\int\int \mathrm{d}^3x\, n^2(\boldsymbol{r}) \left[\mathcal{E}_{\mathrm{DR}}(\boldsymbol{K}^{(+)}, \mu, \boldsymbol{r})\right]^* \cdot \mathcal{E}_{\mathrm{DL}}(\boldsymbol{k}'^{(-)}, \mu', \boldsymbol{r})$$

$$= 0 . \tag{122}$$

5.3   Electric Field Operator in Half-Space Problems

Based on the detector modes, the electric field operator is described by

$$\hat{\boldsymbol{E}}(\boldsymbol{r}, t) = \sum_{\mu=1}^{2} \int\int\int_{0 \leq s_\parallel < 1} \mathrm{d}^3K^{(+)} \left[\frac{\hbar K}{(2\pi)^3\epsilon_0}\right]^{1/2}$$

$$\times [\hat{a}_{\mathrm{DR}}(\boldsymbol{K}^{(+)}, \mu)\mathcal{E}_{\mathrm{DR}}(\boldsymbol{K}^{(+)}, \mu, \boldsymbol{r})e^{-iKt} + \mathrm{H.c.}]$$

$$+ \sum_{\mu=1}^{2} \int\int\int_{0 \leq \kappa_\parallel < 1} \mathrm{d}^3k^{(-)} \left[\frac{\hbar K}{(2\pi)^3\epsilon_0}\right]^{1/2}$$

$$\times [\hat{a}_{\mathrm{DL}}(\boldsymbol{k}^{(-)}, \mu)\mathcal{E}_{\mathrm{DL}}(\boldsymbol{k}^{(-)}, \mu, \boldsymbol{r})e^{-iKt} + \mathrm{H.c.}] , \tag{123}$$

where $\hat{a}_{\text{DR}}(\boldsymbol{K}^{(+)}, \mu)$, $\hat{a}_{\text{DR}}^{\dagger}(\boldsymbol{K}^{(+)}, \mu)$ and $\hat{a}_{\text{DL}}(\boldsymbol{k}^{(-)}, \mu)$, $\hat{a}_{\text{DL}}^{\dagger}(\boldsymbol{k}^{(-)}, \mu)$ are the annihilation and creation operators of photons characterized by wavevector $\boldsymbol{K}^{(+)}$ or $\boldsymbol{k}^{(-)}$ for each polarization state μ. The commutation relations between these operators are given as follows;

$$\left[\hat{a}_{\text{DR}}(\boldsymbol{K}^{(+)}, \mu), \hat{a}_{\text{DR}}^{\dagger}(\boldsymbol{K}'^{(+)}, \mu')\right] = \delta_{\mu\mu'}\delta(\boldsymbol{K}^{(+)} - \boldsymbol{K}'^{(+)}) \,, \tag{124}$$

$$\left[\hat{a}_{\text{DL}}(\boldsymbol{k}^{(-)}, \mu), \hat{a}_{\text{DL}}^{\dagger}(\boldsymbol{k}'^{(-)}, \mu')\right] = \delta_{\mu\mu'}\delta(\boldsymbol{k}^{(-)} - \boldsymbol{k}'^{(-)}) \,, \tag{125}$$

$$\left[\hat{a}_{\text{DR}}(\boldsymbol{K}^{(+)}, \mu), \hat{a}_{\text{DL}}(\boldsymbol{k}'^{(-)}, \mu')\right] = 0 \,, \tag{126}$$

$$\left[\hat{a}_{\text{DR}}(\boldsymbol{K}^{(+)}, \mu), \hat{a}_{\text{DL}}^{\dagger}(\boldsymbol{k}'^{(-)}, \mu')\right] = 0 \,. \tag{127}$$

Finally, let us consider the Hamiltonian $\hat{\mathcal{H}}$, number operator $\hat{\mathcal{N}}$ and pseudomomentum operator $\hat{\boldsymbol{P}}_{\parallel}$ of the quantized electromagnetic field. We obtain

$$\hat{\mathcal{H}} = \sum_{\mu=1}^{2} \int \int \int_{0 \leq s_{\parallel} < 1} \mathrm{d}^3 K^{(+)} \hbar K \hat{a}_{\text{DR}}^{\dagger}(\boldsymbol{K}^{(+)}, \mu) \hat{a}_{\text{DR}}(\boldsymbol{K}^{(+)}, \mu)$$
$$+ \sum_{\mu=1}^{2} \int \int \int_{0 \leq \kappa_{\parallel} < 1} \mathrm{d}^3 k^{(-)} \hbar K \hat{a}_{\text{DL}}^{\dagger}(\boldsymbol{k}^{(-)}, \mu) \hat{a}_{\text{DL}}(\boldsymbol{k}^{(-)}, \mu) \,, \tag{128}$$

$$\hat{\mathcal{N}} = \sum_{\mu=1}^{2} \int \int \int_{0 \leq s_{\parallel} < 1} \mathrm{d}^3 K^{(+)} \hat{a}_{\text{DR}}^{\dagger}(\boldsymbol{K}^{(+)}, \mu) \hat{a}_{\text{DR}}(\boldsymbol{K}^{(+)}, \mu)$$
$$+ \sum_{\mu=1}^{2} \int \int \int_{0 \leq \kappa_{\parallel} < 1} \mathrm{d}^3 k^{(-)} \hat{a}_{\text{DL}}^{\dagger}(\boldsymbol{k}^{(-)}, \mu) \hat{a}_{\text{DL}}(\boldsymbol{k}^{(-)}, \mu) \,, \tag{129}$$

$$\hat{\boldsymbol{P}}_{\parallel} = \sum_{\mu=1}^{2} \int \int \int_{0 \leq s_{\parallel} < 1} \mathrm{d}^3 K^{(+)} \hbar K \hat{\boldsymbol{s}}_{\parallel} \hat{a}_{\text{DR}}^{\dagger}(\boldsymbol{K}^{(+)}, \mu) \hat{a}_{\text{DR}}(\boldsymbol{K}^{(+)}, \mu)$$
$$+ \sum_{\mu=1}^{2} \int \int \int_{0 \leq \kappa_{\parallel} < 1} \mathrm{d}^3 k^{(-)} \hbar K \hat{\boldsymbol{s}}_{\parallel} \hat{a}_{\text{DL}}^{\dagger}(\boldsymbol{k}^{(-)}, \mu) \hat{a}_{\text{DL}}(\boldsymbol{k}^{(-)}, \mu) \,. \tag{130}$$

The photon-number state corresponding to the mode specified by $\boldsymbol{K}^{(+)}$ and μ or $\boldsymbol{k}^{(-)}$ and μ, is generated by operating $\hat{a}_{\text{DR}}^{\dagger}(\boldsymbol{K}^{(+)}, \mu)$ or $\hat{a}_{\text{DL}}^{\dagger}(\boldsymbol{k}^{(-)}, \mu)$ to the vacuum state $|0\rangle$ as

$$\left| D, N(\boldsymbol{K}^{(+)}, \mu) \right\rangle = (1/\sqrt{N!})[\hat{a}_{\text{DR}}^{\dagger}(\boldsymbol{K}^{(+)}, \mu)]^N |0\rangle \,, \tag{131}$$

$$\left| D, N(\boldsymbol{k}^{(-)}, \mu) \right\rangle = (1/\sqrt{N!})[\hat{a}_{\text{DL}}^{\dagger}(\boldsymbol{k}^{(-)}, \mu)]^N |0\rangle \,, \tag{132}$$

where D indicates that the state is defined on the basis of the detector modes. When we consider a single-photon emission process of atoms near a planar dielectric surface, the transition amplitudes are given, in a first-order approximation, by the nonzero matrix elements of the creation operators, $\hat{a}_{\mathrm{DR}}^{\dagger}(\boldsymbol{K}^{(+)}, \mu)$ and $\hat{a}_{\mathrm{DL}}^{\dagger}(\boldsymbol{k}^{(-)}, \mu)$, as follows;

$$\left\langle D, N(\boldsymbol{K}^{(+)}, \mu) + 1 \middle| \hat{a}_{\mathrm{DR}}^{\dagger}(\boldsymbol{K}'^{(+)}, \mu') \middle| D, N(\boldsymbol{K}'^{(+)}, \mu') \right\rangle$$
$$= \sqrt{(N+1)} \delta_{\mu\mu'} \delta(\boldsymbol{K}^{(+)} - \boldsymbol{K}'^{(+)}) , \quad (133)$$

$$\left\langle D, N(\boldsymbol{k}^{(-)}, \mu) + 1 \middle| \hat{a}_{\mathrm{DL}}^{\dagger}(\boldsymbol{k}'^{(-)}, \mu') \middle| D, N(\boldsymbol{k}'^{(-)}, \mu') \right\rangle$$
$$= \sqrt{(N+1)} \delta_{\mu\mu'} \delta(\boldsymbol{k}^{(-)} - \boldsymbol{k}'^{(-)}) . \quad (134)$$

5.4   Spontaneous Emission into Right Half-Space

We consider the photon-emission process from an excited atom placed in the right half-space (vacuum side) near a planar dielectric surface. The interaction Hamiltonian between the atom and the electromagnetic field is given by

$$\hat{V}(t) = -\frac{e}{m_{\mathrm{e}}} \hat{\boldsymbol{A}}(\boldsymbol{r}_0 + \boldsymbol{R}, t) \cdot \boldsymbol{p}_0 , \quad (135)$$

where the atomic-position vector is $\boldsymbol{R} = (X, Y, Z)$ $(Z > 0)$, $\boldsymbol{r}_0 = (x_0, y_0, z_0)$ the relative position vector of the atomic electron with respect to the nucleus, and e, m_{e}, and $\boldsymbol{p}_0$ the electron charge, mass, and momentum, respectively. $\hat{\boldsymbol{A}}$ is the vector potential in the Coulomb gauge obtained from (123) by using the relation

$$\hat{\boldsymbol{E}}(\boldsymbol{r}, t) = -\frac{\partial}{\partial t} \hat{\boldsymbol{A}}(\boldsymbol{r}, t) . \quad (136)$$

We will consider a photon-emission process of a two-level atom with a μ-polarized outgoing wave in the vacuum-side half-space with wavevector $\boldsymbol{K}^{(+)} = K(s_x, s_y, s_z)$. Here it is stressed that a single detector mode is specified by its outgoing wavevector $\boldsymbol{K}^{(+)}$ and polarization μ. For the photon emission in the mode specified by $\boldsymbol{K}^{(+)}$ and μ, the final state is described by $|f\rangle = |D, 1(\boldsymbol{K}^{(+)}, \mu)\rangle |\varphi_f\rangle$, where $|\varphi_f\rangle$ corresponds to the atomic ground state. The initial state of the system is described by $|i\rangle = |0\rangle |\varphi_i\rangle$, where $|\varphi_i\rangle$ corresponds to an atomic excited state under consideration. By separating the temporal evolution of the wave functions, the matrix element of the operator in (135) can be represented by using the time-independent matrix element V_{fi} as

$$V_{fi}(t) = V_{fi} e^{-\mathrm{i}(\omega_0 - K)t} , \quad (137)$$

where ω_0 is the atomic-transition frequency between the initial and final states. The probability $\mathrm{d}\Gamma$ for the $i \to f$ transition resulting in a single-photon emission is given by

$$\mathrm{d}\Gamma = \frac{2\pi}{\hbar^2} |V_{fi}|^2 \delta(\omega_0 - K) \mathrm{d}\rho(K) , \quad (138)$$

where $d\rho(K)$ indicates the final-state mode density of the electromagnetic field.

When we consider the single-photon emission into the vacuum side, the state of the photon is described by the R-detector mode alone. From (123), (133), and (135), we can find the time-independent transition matrix element as

$$V_{fi}(\boldsymbol{K}^{(+)}, \mu) = -\mathrm{i}\left(\frac{e}{m_e}\right)\left[\frac{\hbar}{(2\pi)^3 K \epsilon_0}\right]^{1/2}$$
$$\times \langle\varphi_f|\left[\boldsymbol{\mathcal{E}}_{\mathrm{DR}}(\boldsymbol{K}^{(+)}, \mu, \boldsymbol{r}_0 + \boldsymbol{R})\right]^* \cdot \boldsymbol{p}_0 |\varphi_i\rangle .\tag{139}$$

Using the long-wavelength approximation and the relation $\langle\varphi_f|\boldsymbol{p}_0|\varphi_i\rangle = -\mathrm{i}m_e\omega_0\langle\varphi_f|\boldsymbol{r}_0|\varphi_i\rangle$, the matrix element can be written in the following form:

$$V_{fi}(\boldsymbol{K}^{(+)}, \mu) = V_{fi}^{(\mathrm{I})}(\boldsymbol{K}^{(+)}, \mu) + V_{fi}^{(\mathrm{R})}(\boldsymbol{K}^{(-)}, \mu) ,\tag{140}$$

where

$$V_{fi}^{(\mathrm{I})}(\boldsymbol{K}^{(+)}, \mu) = -\omega_0\left[\frac{\hbar}{2(2\pi)^3 K \epsilon_0}\right]^{1/2}$$
$$\times \left[\boldsymbol{\varepsilon}(\hat{\boldsymbol{s}}^{(+)}, \mu) \cdot \boldsymbol{d}_{fi}\right]\exp(-\mathrm{i}K\hat{\boldsymbol{s}}^{(+)} \cdot \boldsymbol{R}) ,\tag{141}$$

$$V_{fi}^{(\mathrm{R})}(\boldsymbol{K}^{(-)}, \mu) = -\omega_0\left[\frac{\hbar}{2(2\pi)^3 K \epsilon_0}\right]^{1/2}$$
$$\times \left[\boldsymbol{\varepsilon}(\hat{\boldsymbol{s}}^{(-)}, \mu) \cdot \boldsymbol{d}_{fi}\right]R_{\mathrm{R}}(s_z, \mu)\exp(-\mathrm{i}K\hat{\boldsymbol{s}}^{(-)} \cdot \boldsymbol{R}) .\tag{142}$$

Here, we denote $\boldsymbol{d}_{fi} = e\langle\varphi_f|\boldsymbol{r}_0|\varphi_i\rangle$. $V_{fi}^{(\mathrm{I})}$ indicates the matrix element of interaction between $\boldsymbol{\mathcal{E}}_{\mathrm{DR}}^{(\mathrm{I})}$ and the electric dipole. $V_{fi}^{(\mathrm{R})}$ indicates the matrix element of interaction between $\boldsymbol{\mathcal{E}}_{\mathrm{DR}}^{(\mathrm{R})}$ and the electric dipole. The radiation field involves only one outgoing wave with wavevector $\boldsymbol{K}^{(+)}$ and polarization μ, so that the mode density $d\rho(K)$ for the final state of the photon for each μ given simply by

$$d\rho(K) = \mathrm{d}^3\boldsymbol{K}^{(+)} = K^2\mathrm{d}K\mathrm{d}\Omega(\hat{\boldsymbol{s}}^{(+)}),\tag{143}$$

with the infinitesimal solid angle $d\Omega(\hat{\boldsymbol{s}}^{(+)}) = \mathrm{d}s_x\mathrm{d}s_y/s_z$ in the direction of unit prapagation vector $\hat{\boldsymbol{s}}^{(+)}$. Here it is stressed that the detector mode provides a clear understanding of the radiation process from the viewpoints of the classical–quantum correspondence and the straightforward evaluation of the final mode density discussed in the above.

Substituting (140) and (143) into (138) and integrating over dK, the differential radiation probability $d\Gamma$ for photon emission into the mode with outgoing wave with $\boldsymbol{K}^{(+)}$ lying in the solid angle $d\Omega(\hat{\boldsymbol{s}}^{(+)})$ is given by

$$\mathrm{d}\Gamma_+(\hat{\boldsymbol{s}}^{(+)}, \mu) = \mathrm{d}\Gamma_+^{(\mathrm{h})}(\hat{\boldsymbol{s}}^{(+)}, \mu) + \mathrm{d}\Gamma_+^{(\mathrm{c})}(\hat{\boldsymbol{s}}^{(+)}, \mu) .\tag{144}$$

Here, $d\Gamma_+^{(h)}$ and $d\Gamma_+^{(c)}$ are defined by

$$d\Gamma_+^{(h)}(\hat{s}^{(+)}, \mu) = \left(\frac{2\pi K^2}{\hbar^2}\right) \left[V_{fi}^{(I)*}(\boldsymbol{K}^{(+)}, \mu) V_{fi}^{(I)}(\boldsymbol{K}^{(+)}, \mu)\right.$$
$$\left. + V_{fi}^{(R)*}(\boldsymbol{K}^{(-)}, \mu) V_{fi}^{(R)}(\boldsymbol{K}^{(-)}, \mu)\right] d\Omega(\hat{s}^{(+)}), \quad (145)$$

$$d\Gamma_+^{(c)}(\hat{s}^{(+)}, \mu) = \left(\frac{2\pi K^2}{\hbar^2}\right) \left[V_{fi}^{(I)*}(\boldsymbol{K}^{(+)}, \mu) V_{fi}^{(R)}(\boldsymbol{K}^{(-)}, \mu)\right.$$
$$\left. + V_{fi}^{(R)*}(\boldsymbol{K}^{(-)}, \mu) V_{fi}^{(I)}(\boldsymbol{K}^{(+)}, \mu)\right] d\Omega(\hat{s}^{(+)}), \quad (146)$$

where $K = \omega_0$. Substituting (141) and (142) into (145) and (146), $d\Gamma_+^{(h)}$ and $d\Gamma_+^{(c)}$ can be obtained as

$$d\Gamma_+^{(h)}(\hat{s}^{(+)}, \mu) = \left(\frac{K^3}{8\pi^2\hbar\epsilon_0}\right) \left\{\left[(\boldsymbol{d}_{fi})^* \cdot \hat{\boldsymbol{\varepsilon}}(\hat{s}^{(+)}, \mu)\right]\left[\hat{\boldsymbol{\varepsilon}}(\hat{s}^{(+)}, \mu) \cdot \boldsymbol{d}_{fi}\right]\right.$$
$$\left. + \left[(\boldsymbol{d}_{fi})^* \cdot \hat{\boldsymbol{\varepsilon}}(\hat{s}^{(-)}, \mu)\right]\left[\hat{\boldsymbol{\varepsilon}}(\hat{s}^{(-)}, \mu) \cdot \boldsymbol{d}_{fi}\right] R_R^2(s_z, \mu)\right\} d\Omega(\hat{s}^{(+)}), (147)$$

$$d\Gamma_+^{(c)}(\hat{s}^{(+)}, \mu) = \left(\frac{K^3}{4\pi^2\hbar\epsilon_0}\right) \Re\left\{\left[(\boldsymbol{d}_{fi})^* \cdot \hat{\boldsymbol{\varepsilon}}(\hat{s}^{(+)}, \mu)\right]\right.$$
$$\left. \times \left[\boldsymbol{\varepsilon}(\hat{s}^{(-)}, \mu) \cdot \boldsymbol{d}_{fi}\right] R_R(s_z, \mu) \exp(2iKs_z Z)\right\} d\Omega(\hat{s}^{(+)}). \quad (148)$$

5.5   Spontaneous Emission into Left Half-Space

Next, let us consider an emission of μ-polarized photon resulting in the outgoing wave in the medium side with $\boldsymbol{k}^{(-)} = nK(\kappa_x, \kappa_y, -\kappa_z)$. The final state corresponds to $|f\rangle = |D, 1(\boldsymbol{k}^{(-)}, \mu)\rangle |\varphi_f\rangle$. From (123), (134), and (135), the time-independent matrix element is given by

$$V_{fi}(\boldsymbol{k}^{(-)}, \mu) = -\left[\frac{\hbar}{(2\pi)^3 K\epsilon_0}\right]^{1/2}$$
$$\times \langle\varphi_f| \left[\boldsymbol{\mathcal{E}}_{DL}(\boldsymbol{k}^{(-)}, \mu, \boldsymbol{R} + \boldsymbol{r}_0)\right]^* \cdot \boldsymbol{p}_0 |\varphi_i\rangle, \quad (149)$$

$$V_{fi}(\boldsymbol{k}^{(-)}, \mu) = V_{fi}^{(T)}(\boldsymbol{K}^{(-)}, \mu), \quad (150)$$

where

$$V_{fi}^{(T)}(\boldsymbol{K}^{(-)}, \mu) = -\frac{\omega_0}{n}\left[\frac{\hbar}{2(2\pi)^3 K\epsilon_0}\right]^{1/2}$$
$$\times \left[\hat{\boldsymbol{\varepsilon}}(\hat{s}^{(-)}, \mu) \cdot \boldsymbol{d}_{fi}\right] T_L(s_z, \mu) \exp(-iK\hat{s}^{(-)} \cdot \boldsymbol{R}). \quad (151)$$

As the final-state mode function is labeled by $\boldsymbol{k}^{(-)}$ and μ in $\boldsymbol{\mathcal{E}}_{DL}(\boldsymbol{k}^{(-)}, \mu, \boldsymbol{r})$, the differential mode density $d\rho(K)$ is given simply by

$$d\rho(K) = d^3\boldsymbol{k}^{(-)} = n^3 K^2 dK d\Omega(\hat{\boldsymbol{\kappa}}^{(-)}), \quad (152)$$

with the solid angle $d\Omega(\hat{\boldsymbol{\kappa}}^{(-)}) = d\kappa_x d\kappa_y/\kappa_z$ in the direction of the unit propagation vector $\hat{\boldsymbol{\kappa}}^{(-)}$. Substituting (151) and (152) into (138) and integrating over dK, the differential transition probability $d\Gamma$ is given by

$$
d\Gamma_-(\hat{\boldsymbol{\kappa}}^{(-)},\mu) = \begin{cases} d\Gamma_-^{(\mathrm{h})}(\hat{\boldsymbol{\kappa}}^{(-)},\mu) & \text{for } 0 \leq \kappa_\| < 1/n \,, \\[2mm] d\Gamma_-^{(\mathrm{t})}(\hat{\boldsymbol{\kappa}}^{(-)},\mu) & \text{for } (1/n) \leq \kappa_\| < 1 \,, \end{cases} \tag{153}
$$

$$
d\Gamma_-^{(\mathrm{h})}(\hat{\boldsymbol{\kappa}}^{(-)},\mu)
$$
$$
= \left(\frac{2\pi n^3 K^2}{\hbar^2}\right) V_{fi}^{(\mathrm{T})*}(\boldsymbol{K}^{(-)},\mu) V_{fi}^{(\mathrm{T})}(\boldsymbol{K}^{(-)},\mu) d\Omega(\hat{\boldsymbol{\kappa}}^{(-)}) \,, \tag{154}
$$

$$
d\Gamma_-^{(\mathrm{t})}(\hat{\boldsymbol{\kappa}}^{(-)},\mu)
$$
$$
= \left(\frac{2\pi n^3 K^2}{\hbar^2}\right) V_{fi}^{(\mathrm{T})*}(\boldsymbol{K}^{(+)},\mu) V_{fi}^{(\mathrm{T})}(\boldsymbol{K}^{(-)},\mu) d\Omega(\hat{\boldsymbol{\kappa}}^{(-)}) \,. \tag{155}
$$

Here, $d\Gamma_-^{(\mathrm{h})}$ and $d\Gamma_-^{(\mathrm{t})}$ indicate the probability of homogeneous photon emission and evanescent photon emission, respectively. Substituting (151) into (154) and (155), respectively, these are obtained as

$$
d\Gamma_-^{(\mathrm{h})}(\hat{\boldsymbol{\kappa}}^{(-)},\mu) = \left(\frac{K^3}{8\pi^2 \hbar \epsilon_0}\right) \left[(\boldsymbol{d}_{fi})^* \cdot \hat{\boldsymbol{\varepsilon}}(\hat{\boldsymbol{s}}^{(-)},\mu)\right] \left[\hat{\boldsymbol{\varepsilon}}(\hat{\boldsymbol{s}}^{(-)},\mu) \cdot \boldsymbol{d}_{fi}\right]
$$
$$
\times n T_{\mathrm{L}}^2(s_z,\mu) d\Omega(\hat{\boldsymbol{\kappa}}^{(-)}) \,, \tag{156}
$$

$$
d\Gamma_-^{(\mathrm{t})}(\hat{\boldsymbol{\kappa}}^{(-)},\mu) = \left(\frac{K^3}{8\pi^2 \hbar \epsilon_0}\right) \left[(\boldsymbol{d}_{fi})^* \cdot \hat{\boldsymbol{\varepsilon}}(\hat{\boldsymbol{s}}^{(+)},\mu)\right] \left[\hat{\boldsymbol{\varepsilon}}(\hat{\boldsymbol{s}}^{(-)},\mu) \cdot \boldsymbol{d}_{fi}\right]
$$
$$
\times n T_{\mathrm{L}}(-s_z,\mu) T_{\mathrm{L}}(s_z,\mu) \exp\left(2\mathrm{i}K s_z Z\right) d\Omega(\hat{\boldsymbol{\kappa}}^{(-)}) \,. \tag{157}
$$

5.6 Radiative Decay Rate and Lifetime of Electric Dipole in Half-Space

The probability of spontaneous photon emission into the right half-space from the atom exerting transition from the initial state $|\varphi_i\rangle$ to the final state $|\varphi_f\rangle$ is obtained by integrating $d\Gamma_+$ over $\mathcal{A}_+$, and summing up $d\Gamma_+$ for all the possible polarizations of radiation, μ. The result is given by the sum of two components,

$$
\Gamma_+ = \Gamma_+^{(\mathrm{h})} + \Gamma_+^{(\mathrm{c})} \,, \tag{158}
$$

where $\Gamma_+^{(\mathrm{h})}$ and $\Gamma_+^{(\mathrm{c})}$ are calculated, respectively, by integrating (147) and (148) over ds_x and ds_y $(0 \leq s_\| < 1)$;

$$\Gamma_+^{(h)} = \left(\frac{K^3}{8\pi^2\hbar\epsilon_0}\right)\sum_{\mu=1}^{2}\int\!\!\int_{0\leq s_{\parallel}<1} ds_x ds_y \frac{1}{s_z}$$

$$\times\left\{\left[(\boldsymbol{d}_{fi})^*\cdot\hat{\boldsymbol{\varepsilon}}(\hat{\boldsymbol{s}}^{(+)},\mu)\right]\left[\hat{\boldsymbol{\varepsilon}}(\hat{\boldsymbol{s}}^{(+)},\mu)\cdot\boldsymbol{d}_{fi}\right]\right.$$

$$\left.+\left[(\boldsymbol{d}_{fi})^*\cdot\hat{\boldsymbol{\varepsilon}}(\hat{\boldsymbol{s}}^{(-)},\mu)\right]\left[\hat{\boldsymbol{\varepsilon}}(\hat{\boldsymbol{s}}^{(-)},\mu)\cdot\boldsymbol{d}_{fi}\right]R_R^2(s_z,\mu)\right\}\ ,\quad (159)$$

$$\Gamma_+^{(c)} = \left(\frac{K^3}{4\pi^2\hbar\epsilon_0}\right)\Re\left\{\sum_{\mu=1}^{2}\int\!\!\int_{0\leq s_{\parallel}<1} ds_x ds_y \frac{1}{s_z}\right.$$

$$\left.\times\left[(\boldsymbol{d}_{fi})^*\cdot\hat{\boldsymbol{\varepsilon}}(\hat{\boldsymbol{s}}^{(+)},\mu)\right]\left[\boldsymbol{\varepsilon}(\hat{\boldsymbol{s}}^{(-)},\mu)\cdot\boldsymbol{d}_{fi}\right]R_R(s_z,\mu)\exp\left(2\mathrm{i}Ks_zZ\right)\right\}\ .\quad (160)$$

On the other hand, the probability of spontaneous emission into the left half-space is obtained by integrating $d\Gamma_-$ over $\mathcal{A}_-$, and summing up $d\Gamma_-$ for all the possible polarizations, μ. The result is given also as the sum of two components,

$$\Gamma_- = \Gamma_-^{(h)} + \Gamma_-^{(t)}\ ,\quad\quad\quad\quad (161)$$

where $\Gamma_-^{(h)}$ and $\Gamma_-^{(t)}$ are calculated by integrating (156) and (157);

$$\Gamma_-^{(h)} = \left(\frac{K^3}{8\pi^2\hbar\epsilon_0}\right)\sum_{\mu=1}^{2}\int\!\!\int_{0\leq\kappa_{\parallel}<1/n} d\kappa_x d\kappa_y \frac{1}{\kappa_z}$$

$$\times\left[(\boldsymbol{d}_{fi})^*\cdot\hat{\boldsymbol{\varepsilon}}(\hat{\boldsymbol{s}}^{(-)},\mu)\right]\left[\hat{\boldsymbol{\varepsilon}}(\hat{\boldsymbol{s}}^{(-)},\mu)\cdot\boldsymbol{d}_{fi}\right]nT_L^2(s_z,\mu)\ ,\quad (162)$$

$$\Gamma_-^{(t)} = \left(\frac{K^3}{8\pi^2\hbar\epsilon_0}\right)\sum_{\mu=1}^{2}\int\!\!\int_{(1/n)\leq\kappa_{\parallel}<n} d\kappa_x d\kappa_y \frac{1}{\kappa_z}$$

$$\times\left[(\boldsymbol{d}_{fi})^*\cdot\hat{\boldsymbol{\varepsilon}}(\hat{\boldsymbol{s}}^{(+)},\mu)\right]\left[\hat{\boldsymbol{\varepsilon}}(\hat{\boldsymbol{s}}^{(-)},\mu)\cdot\boldsymbol{d}_{fi}\right]$$

$$\times nT_L(-s_z,\mu)T_L(s_z,\mu)\exp\left(2\mathrm{i}Ks_zZ\right)\ .\quad (163)$$

These results can be transformed, respectively, to the integrals with respect to ds_x and ds_y with the use of (68) and (69). The results are as follows;

$$\Gamma_-^{(h)} = \left(\frac{K^3}{8\pi^2\hbar\epsilon_0}\right)\sum_{\mu=1}^{2}\int\!\!\int_{0\leq s_{\parallel}<1} ds_x ds_y \frac{1}{s_z}$$

$$\times\left[(\boldsymbol{d}_{fi})^*\cdot\hat{\boldsymbol{\varepsilon}}(\hat{\boldsymbol{s}}^{(-)},\mu)\right]\left[\hat{\boldsymbol{\varepsilon}}(\hat{\boldsymbol{s}}^{(-)},\mu)\cdot\boldsymbol{d}_{fi}\right]\left[1-R_R^2(s_z,\mu)\right]\ ,\quad (164)$$

$$\Gamma_-^{(t)} = \left(\frac{K^3}{4\pi^2\hbar\epsilon_0}\right)\Re\left\{\sum_{\mu=1}^{2}\int\!\!\int_{1\leq s_{\parallel}<n} ds_x ds_y \frac{1}{s_z}\right.$$

$$\left.\times\left[(\boldsymbol{d}_{fi})^*\cdot\hat{\boldsymbol{\varepsilon}}(\hat{\boldsymbol{s}}^{(+)},\mu)\right]\left[\hat{\boldsymbol{\varepsilon}}(\hat{\boldsymbol{s}}^{(-)},\mu)\cdot\boldsymbol{d}_{fi}\right]R_R(s_z,\mu)\exp\left(2\mathrm{i}Ks_zZ\right)\right\}\ ,\quad (165)$$

where we have used (100), (103), and the relation between T_{L} and T_{R},

$$T_{\mathrm{L}}(s_z, \mu) = \left(\frac{n\kappa_z}{s_z}\right) T_{\mathrm{R}}(s_z, \mu) . \tag{166}$$

Since the radiated power per second, I, is related to the photon-emission probability, or the radiative decay rate of the excited atom as $I = \hbar K\Gamma$, the results with the full quantum treatment given in (159), (160), (164) and (165) can be related, respectively, to the results of classical treatment as

$$\hbar K\Gamma_\pm^{(\mathrm{h})} \to I_\pm^{(\mathrm{h})}, \quad \hbar K\Gamma_+^{(\mathrm{c})} \to I_+^{(\mathrm{c})}, \quad \hbar K\Gamma_-^{(\mathrm{t})} \to I_-^{(\mathrm{t})} .$$

Here, it is understood that $\Gamma_+^{(\mathrm{h})} + \Gamma_-^{(\mathrm{h})}$ corresponds to the probability of the spontaneous emission, or the spontaneous decay rate of the atomic excited state, in vacuum, because $I_+^{(\mathrm{h})} + I_-^{(\mathrm{h})}$ gives the total power per second radiated from the excited atom in vacuum as shown in Sect. 4.5. There arise additional terms in the spontaneous decay rate in the half-space problem; $\Gamma_+^{(\mathrm{c})}$ is due to the interference between the direct and reflected waves propagating in the right half-space, and $\Gamma_-^{(\mathrm{t})}$ corresponds to the tunneling energy transport from the excited atom to the dielectric medium via evanescent waves. It is stressed that the latter shows one of the most important results of the optical near-field interactions of atoms with a dielectric surface. Therefore, the probability of a photon emission, or the spontaneous decay rate, is represented as the sum of the free-space term and the half-space correction by

$$\Gamma = \Gamma_+ + \Gamma_- = \Gamma^{(0)} + \Delta\Gamma , \tag{167}$$

where

$$\Gamma^{(0)} = \Gamma_+^{(\mathrm{h})} + \Gamma_-^{(\mathrm{h})} , \tag{168}$$

$$\Delta\Gamma = \Gamma_+^{(\mathrm{c})} + \Gamma_-^{(\mathrm{t})} . \tag{169}$$

$\Gamma^{(0)}$ indicates the probability of photon emission in vacuum, and $\Delta\Gamma$ its half-space correction due to atom–surface interactions. $\Delta\Gamma$, in turn, is composed of $\Gamma_+^{(\mathrm{c})}$ as the correction due to interference and $\Gamma_-^{(\mathrm{t})}$ as that due to the tunneling. Substituting (159) and (164) into (168) and replacing the variables of integration $\mathrm{d}s_x$ and $\mathrm{d}s_y$ by α and β according to $s_x = \sin\alpha\cos\beta$ and $s_y = \sin\alpha\sin\beta$, we obtain the angular-spectrum representation of $\Gamma^{(0)}$ as

$$\Gamma^{(0)} = \left(\frac{K^3}{8\pi^2\hbar\epsilon_0}\right) \sum_{\mu=1}^{2} \int_0^\pi \sin\alpha\,\mathrm{d}\alpha \int_0^{2\pi} \mathrm{d}\beta$$
$$\times \left[(\boldsymbol{d}_{fi})^* \cdot \hat{\boldsymbol{\varepsilon}}(\hat{\boldsymbol{s}}^{(+)}, \mu)\right] \left[\hat{\boldsymbol{\varepsilon}}(\hat{\boldsymbol{s}}^{(+)}, \mu) \cdot \boldsymbol{d}_{fi}\right] . \tag{170}$$

In the same way, we obtain the angular-spectrum representation of (160) and (165), respectively, as follows;

$$
\Gamma_+^{(c)} = \left(\frac{K^3}{4\pi^2\hbar\epsilon_0}\right) \Re\left\{\sum_{\mu=1}^2 \int_0^{\pi/2} \sin\alpha d\alpha \int_0^{2\pi} d\beta\right.
$$

$$
\left.\times \left[(\boldsymbol{d}_{fi})^* \cdot \hat{\boldsymbol{\varepsilon}}(\hat{\boldsymbol{s}}^{(+)},\mu)\right]\left[\boldsymbol{\varepsilon}(\hat{\boldsymbol{s}}^{(-)},\mu)\cdot\boldsymbol{d}_{fi}\right]R_R(s_z,\mu)\exp(2\mathrm{i}Ks_zZ)\right\}, \quad (171)
$$

$$
\Gamma_-^{(t)} = \left(\frac{K^3}{4\pi^2\hbar\epsilon_0}\right) \Re\left\{\sum_{\mu=1}^2 \int_{\pi/2}^{\pi/2-\mathrm{i}\gamma_c} \sin\alpha d\alpha \int_0^{2\pi} d\beta\right.
$$

$$
\left.\times \left[(\boldsymbol{d}_{fi})^* \cdot \hat{\boldsymbol{\varepsilon}}(\hat{\boldsymbol{s}}^{(+)},\mu)\right]\left[\boldsymbol{\varepsilon}(\hat{\boldsymbol{s}}^{(-)},\mu)\cdot\boldsymbol{d}_{fi}\right]R_R(s_z,\mu)\exp(2\mathrm{i}Ks_zZ)\right\}. \quad (172)
$$

5.7 Dependence of Radiative Lifetime on Magnetic Quantum Number of Atom in Half-Space Problems

It is useful to describe the electric dipole operator in terms of spherical basis $\hat{\boldsymbol{e}}_q$ ($q = \pm 1, 0$), as

$$
\boldsymbol{d} = \sum_{q=-1}^{+1} (-1)^q d_q^{(1)} \hat{\boldsymbol{e}}_{-q}. \quad (173)
$$

With these expansion coefficients, the transition matrix element of the electric dipole moment of atom, $\boldsymbol{d}_{fi}$, is represented as

$$
\boldsymbol{d}_{fi} = \sum_{q=-1}^{+1} (-1)^q \langle f| d_q^{(1)} |i\rangle \hat{\boldsymbol{e}}_{-q}. \quad (174)
$$

Substituting this into (170) and using the representations of the polarization vectors, $\hat{\boldsymbol{\varepsilon}}(\hat{\boldsymbol{s}}^{(+)},\mu)$, given in (48) and (49), we obtain

$$
\Gamma^{(0)} = \left(\frac{K^3}{3\pi\hbar\epsilon_0}\right)\sum_{q=-1}^{+1}\left|\langle f| d_q^{(1)} |i\rangle\right|^2 = \frac{K^3|\boldsymbol{d}_{fi}|^2}{3\pi\hbar\epsilon_0}. \quad (175)
$$

Summing up $\Gamma^{(0)}$ over all the possible atomic final states, we obtain the radiative decay rate of the atomic initial state $|\varphi_i\rangle$ being equivalent to the classical result given in (35). Substititing (174) into (171) and (172) and using the representations of the polarization vectors, $\hat{\boldsymbol{\varepsilon}}(\hat{\boldsymbol{s}}^{(\pm)},\mu)$, we obtain the half-space corrections of the radiative decay rate as

$$
\Gamma_+^{(c)} = \left(\frac{K^3}{2\pi\hbar\epsilon_0}\right)\left[\left(\left|\langle f| d_{+1}^{(1)} |i\rangle\right|^2 + \left|\langle f| d_{-1}^{(1)} |i\rangle\right|^2\right)I_1^{(c)}\right.
$$

$$
\left. + \left|\langle f| d_0^{(1)} |i\rangle\right|^2 I_0^{(c)}\right], \quad (176)
$$

$$\Gamma_{-}^{(t)} = \left(\frac{K^3}{2\pi\hbar\epsilon_0}\right)\left[\left(\left|\langle f|\, d_{+1}^{(1)}\,|i\rangle\right|^2 + \left|\langle f|\, d_{-1}^{(1)}\,|i\rangle\right|^2\right) I_1^{(t)}\right.$$

$$\left. + \left|\langle f|\, d_0^{(1)}\,|i\rangle\right|^2 I_0^{(t)}\right] ,\qquad (177)$$

where

$$I_1^{(c)} = \int_0^1 ds_z \frac{1}{2}\left[R_R(s_z,1) - s_z^2 R_R(s_z,2)\right]\cos(2Ks_zZ) ,\qquad (178)$$

$$I_0^{(c)} = \int_0^1 ds_z(1-s_z^2)R_R(s_z,2)\cos(2Ks_zZ) ,\qquad (179)$$

$$I_1^{(t)} = \int_0^{\sqrt{n^2-1}} d\xi_z \frac{1}{2}\left[\Im m\left\{R_R(i\xi_z,1)\right\}\right.$$

$$\left. + \xi_z^2 \Im m\left\{R_R(i\xi_z,2)\right\}\right]\exp\left(-2K\xi_zZ\right) ,\qquad (180)$$

$$I_0^{(t)} = \int_0^{\sqrt{n^2-1}} d\xi_z \left(1+\xi_z^2\right)\Im m\left\{R_R(i\xi_z,2)\right\}\exp\left(-2K\xi_zZ\right) .\qquad (181)$$

Here, we consider the initial state as an atomic excited state without spin described by

$$|\varphi_i\rangle = |n_i,\ell_i,m_i\rangle ,\qquad (182)$$

where ℓ_i indicates the angular momentum, m_i the magnetic quantum number, and n_i the other quantum number characterizing the atomic excited state. We also describe the final state as

$$|\varphi_f\rangle = |n_f,\ell_f,m_f\rangle .\qquad (183)$$

According to the Wigner–Eckart theorem, the matrix element of the electric dipole operator can be evaluated as

$$\left|\langle n_f,\ell_f,m_f|\, d_q^{(1)}\,|n_i,\ell_i,m_i\rangle\right|^2$$

$$= \langle \ell_i,1;m_i,q|\,\ell_i,1;\ell_f,m_f\rangle^2\frac{\left|\langle n_f,\ell_f||d^{(1)}||n_i,\ell_i\rangle\right|^2}{2\ell_f+1} ,\qquad (184)$$

where $\langle \ell_i,1;m_i,q|\,\ell_i,1;\ell_f,m_f\rangle$ indicates Clebsch–Gordan coefficient [30]. Summing up the probability over all the possible values of m_f for a given m_i, we have the total probability of photon emission at a given frequency from the initial state of the atom. For an atom in free space the result is given by

$$\Gamma^{(0)}(n_i,\ell_i \to n_f,\ell_f) = \left(\frac{K^3}{3\pi\hbar\epsilon_0}\right)\frac{\left|\langle n_f,\ell_f||d^{(1)}||n_i,\ell_i\rangle\right|^2}{(2\ell_i+1)} .\qquad (185)$$

It is noted that the $\Gamma^{(0)}$ is independent of the initial value of m_i because of spatial isotropy. The summation is carried out by using the formula;

$$\sum_{m_f}\sum_q \left| \langle n_f, \ell_f, m_f| \, d_q^{(1)} \, |n_i, \ell_i, m_i\rangle \right|^2$$

$$= \frac{1}{2\ell_i + 1} \left| \langle n_f, \ell_f ||d^{(1)}|| n_i, \ell_i \rangle \right|^2 . \qquad (186)$$

On the other hand, summing up the probability of (176) and (177) over all the possible values of m_f for a given m_i, we obtain for the half-space problem;

$$\Gamma_+^{(c)}(n_i, \ell_i, m_i \to n_f, \ell_f) = \frac{3}{2}\Gamma^0(n_i, \ell_i \to n_f, \ell_f)$$

$$\times \left\{ I_1^{(c)} + \left(\frac{2\ell_i + 1}{2\ell_f + 1} \right) \langle \ell_i, 1; m_i, 0| \, \ell_i, 1; \ell_f, m_i \rangle^2 \left(I_0^{(c)} - I_1^{(c)} \right) \right\} , \qquad (187)$$

$$\Gamma_-^{(t)}(n_i, \ell_i, m_i \to n_f, \ell_f) = \frac{3}{2}\Gamma^0(n_i, \ell_i \to n_f, \ell_f)$$

$$\times \left\{ I_1^{(t)} + \left(\frac{2\ell_i + 1}{2\ell_f + 1} \right) \langle \ell_i, 1; m_i, 0| \, \ell_i, 1; \ell_f, m_i \rangle^2 \left(I_0^{(t)} - I_1^{(t)} \right) \right\} . \qquad (188)$$

It is emphasized that these results depend on the initial value of m_i because spatial anisotropy is introduced in half-space problems.

As a special case, we will consider photon emission associated with an atomic transition into the final state with $\ell_f = \ell_i - 1$. The resulting $\Gamma_+^{(c)}$ and $\Gamma_-^{(t)}$ involve the coefficient

$$\langle \ell_i, 1; m_i, 0| \, \ell_i, 1; \ell_i - 1, m_i \rangle = -\sqrt{\frac{(\ell_i + m_i)(\ell_i - m_i)}{\ell_i(2\ell_i + 1)}} . \qquad (189)$$

As a useful example, we show the results for $\Gamma_+^{(c)}$ and $\Gamma_-^{(t)}$ corresponding to the atomic transition from the initial state with $\ell_i = 1$, $m_i = \pm 1$ to the final state with $\ell_f = 0$;

$$\Gamma_+^{(c)}(n_i, 1, \pm 1 \to n_f, 0) = \frac{3}{2}\Gamma^0(n_i, 1 \to n_f, 0)$$

$$\times \int_0^1 ds_z \frac{1}{2} \left[R_R(s_z, 1) - s_z^2 R_R(s_z, 2) \right] \cos\left(2K s_z Z\right) , \qquad (190)$$

$$\Gamma_-^{(t)}(n_i, 1, \pm 1 \to n_f, 0) = \frac{3}{2}\Gamma^0(n_i, 1 \to n_f, 0)$$

$$\times \int_0^{\sqrt{n^2-1}} d\xi_z \frac{1}{2} \left[\Im\{R_R(i\xi_z, 1)\} \right.$$

$$\left. + \xi_z^2 \Im\{R_R(i\xi_z, 2)\} \right] \exp\left(-2K\xi_z Z\right) . \qquad (191)$$

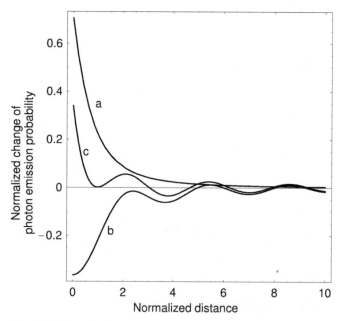

Fig. 20. Normalized spontaneous emission probability calculated as the function of the normalized distance KZ between the dipole and the dielectric surface for the atomic transition from the initial state with $\ell_i = 1$, $m_i = \pm1$ to the final state with $\ell_f = 0$. (a) $\Gamma_-^{(t)}$ normalized by $\Gamma^0(n_i, 1 \to n_f, 0)$ corresponding to the interaction with evanescent waves involved in the L-detector mode. (b) $\Gamma_-^{(c)}$ normalized by Γ^0 corresponding to the interaction with homogeneouse waves involved in the R-detector mode. (c) $\Delta\Gamma$ normalized by Γ^0

For the atomic transition from the initial state with $\ell_i = 1$, $m_i = 0$ to the final state with $\ell_f = 0$, we obtain

$$\Gamma_+^{(c)}(n_i, 1, 0 \to n_f, 0) = \frac{3}{2}\Gamma^0(n_i, 1 \to n_f, 0)$$

$$\times \int_0^1 ds_z \left(1 - s_z^2\right) R_R(s_z, 2) \cos\left(2K s_z Z\right) , \quad (192)$$

$$\Gamma_-^{(t)}(n_i, 1, 0 \to n_f, 0) = \frac{3}{2}\Gamma^0(n_i, 1 \to n_f, 0)$$

$$\times \int_0^{\sqrt{n^2-1}} d\xi_z \left(1 + \xi_z^2\right) \Im m\left\{R_R(i\xi_z, 2)\right\} \exp\left(-2K\xi_z Z\right) . \quad (193)$$

We show in Fig. 20 the numerical results of $\Gamma_+^{(c)}$ and $\Gamma_-^{(t)}$ as a function of the normalized distance KZ between the atom and the dielectric surface. The results are calculated for the atomic transition from the initial state with $\ell_i = 1$ and $m_i = \pm1$ to the final state with $\ell_f = 0$. This corresponds to the

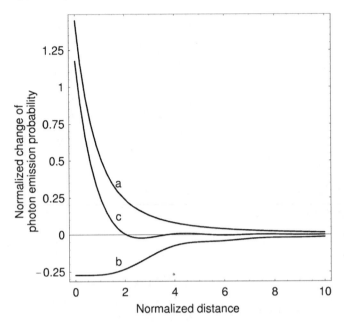

Fig. 21. Normalized spontaneous emission probability as a function of the normalized distance KZ calculated for the atomic transition from the initial state with $\ell_i = 1$, $m_i = 0$ to the final state with $\ell_f = 0$. The curves a, b, c indicate, respectively, $\Gamma_-^{(t)}$, $\Gamma_+^{(c)}$, and $\Gamma_-^{(t)} + \Gamma_+^{(c)}$, normalized by Γ^0

orientation of the classical point dipole in the direction parallel to the planar boundary. The refractive index of the medium is assumed as $n = 1.45$. The spontaneous decay rates are normalized by the probability of spontaneous emission in free space (vacuum), $\Gamma^0(n_i, 1 \to n_f, 0)$. The curve a corresponds to the interaction between the atomic dipole and the evanescent wave involved in the L-detector mode. This process corresponds to the tunnnering of excitation due to the coupling via evanescent waves. The curve b corresponds to the interaction between the atomic dipole and the homogeneous wave involved in the R-detector mode. This process showing an oscillation of the decay rate corresponds to the interference between the direct radiation of the dipole into the right half-space and the reflected radiation of the dipole from the dielectric surface. The curve c corresponds to $\Delta\Gamma$ given as the sum of the results shown in the curves a and b. The results show that the normalized $\Gamma_-^{(t)}$ is strongly enhanced in the near-field regime.

Figure 21 shows the results calculated for the atomic transition from the initial state with $\ell_i = 1$ and $m_i = 0$ to the final state with $\ell_f = 0$. This corresponds to the orientation of the classical point dipole in the direction perpendicular to the planar boundary. The refractive index of the medium is also $n = 1.45$. The normalized value of $\Gamma_-^{(t)}$ for $m_i = 0$ is larger than that calculated for $m_i = \pm 1$.

These results are especially useful when we investigate the spin polarization or manipulation of atoms by means of optical excitation and spontaneous emission near a dielectric surface. Such a process in free space is well known as optical pumping, which produces atomic spin polarization by means of a repeated cycle of optical excitation with circularly polarized light followed by spontaneous emission into isotropic space. The optical pumping in free space corresponds to a transfer of angular momentum from optical fields to atoms. The optical pumping in the near-field regime corresponds to a transfer of angular pseudomomentum of optical fields to atoms.

6 Quantum Theory of Multipole Radiation in Optical Near-Field Regime

In this section we extend our study into quantum optical theory of multipole radiation in optical near fields based on the detector mode we have developed in the previous sections for half-space problems. One of the most important results derived in the following is that multipole radiations are strongly enhanced when they are placed in the optical near field of matter. Indeed the near-field enhancement of the radiative decay rate of electric dipoles investigated in the previous sections is a very interesting and remarkable property from the viewpoints of both the tunneling of optical excitation and cavity QED, but the basic characteristics are still under the influence of ordinary radiation of homogeneous waves. In contrast, multipole processes in free space involve very small amplitude of radiation into far field due to interaction with propagating waves, so that an optical near-field enhancement of the multipole radiation plays the dominant role in near-field interactions of the multipole with its environment without significant loss of excitation energy radiated into far fields. In particular, an electric quadrupole is excited in a mesoscopic electronic system where nonlocal properties play an important role, or in other words a finite spatial extension of the electronic wave function manifests itself in interactions with local electromagnetic fields, its interaction in the near-field regime is as strong as electric-dipole interactions, whereas the radiation loss into far field is still suppressed since the system maintains the nature of the quadrupole when observed from the far-field region. This property is critical if we consider an optical near-field device that exerts its function in terms of optical near-field excitation transfer between electronic two-level systems of mesoscopic nature, such as a pair of quantum dots or molecules. Provided that an electric quadrupole interaction is employed, the local optical near-field coupling in such a device maintains the well-defined meaning even if one considers the system within a restricted space of subwavelength size. Therefore, the study of optical multipole radiation in near field provides an important basis in considerations of nanometer-sized electronic devices in terms of optical near-field interactions.

6.1   Multipole Transition Matrix Elements

In this section, we will consider the spontaneous emission of electric and magnetic multipoles near a planar dielectric surface. Without losing generality, we consider an excited atomic two-level system as an example. Here, we start with the generalized form of electromagnetic interactions, which is basically similar to (135);

$$V_{fi}(t) = -e \int \langle D, 1| \, \hat{A}(r_0 + R, t) \, |0\rangle \cdot j_{fi}(r_0, t) \mathrm{d}^3 x_0 \, , \qquad (194)$$

where $j_{fi}(r_0, t) = (1/m_e) \langle \varphi_f(r_0, t)| \, p_0 \, |\varphi_i(r_0, t)\rangle$ is the transition current associated with the radiative transition in the atomic two-level system from the initial state $|\varphi_i\rangle$ to the final state $|\varphi_f\rangle$.

Employing the half-space system similar to that assumed in the previous section, we consider a single-photon emission process into the half-space of the vacuum side, where the state of the photon radiated is described by using the R detector mode alone. The R detector mode involves only the homogeneous waves, since $0 \le s_\parallel < 1$ corresponding to our assumption that a propagating wave is coupled with a photodetector in the far-field region. The atom placed in the vacuum side at $Z > 0$ interacts with two wave components $\mathcal{E}_{\mathrm{DR}}^{(\mathrm{I})}$ and $\mathcal{E}_{\mathrm{DR}}^{(\mathrm{R})}$ belonging to the R detector mode. We obtain the time-independent matrix element for spontaneous photon emission into the mode with wavevector $K^{(+)}$ and polarization μ in the form of (140), as follows;

$$V_{fi}^{(\mathrm{I})}(K^{(+)}, \mu) = -\mathrm{ie} \left[\frac{\hbar}{2(2\pi)^3 K \epsilon_0} \right]^{1/2}$$
$$\times \left[\hat{\varepsilon}(\hat{s}^{(+)}, \mu) \cdot j_{fi}(K\hat{s}^{(+)}) \right] \exp(-\mathrm{i}K\hat{s}^{(+)} \cdot R) \, , \quad (195)$$

$$V_{fi}^{(\mathrm{R})}(K^{(-)}, \mu) = -\mathrm{ie} \left[\frac{\hbar}{2(2\pi)^3 K \epsilon_0} \right]^{1/2}$$
$$\times \left[\hat{\varepsilon}(\hat{s}^{(-)}, \mu) \cdot j_{fi}(K\hat{s}^{(-)}) \right] R_{\mathrm{R}}(s_z, \mu) \exp(-\mathrm{i}K\hat{s}^{(-)} \cdot R) \, . \quad (196)$$

Here, we have used the Fourier component of the transition current defined by

$$j_{fi}(K\hat{s}^{(\pm)}) = \int \mathrm{d}^3 x_0 \exp(-\mathrm{i}K\hat{s}^{(\pm)} \cdot r_0) j_{fi}(r_0) \, . \qquad (197)$$

For a single-photon emission into the half-space of the medium side, the state of the photon is described by the L detector mode alone. The atom placed at $Z > 0$ in the vacuum side interacts with the component $\mathcal{E}_{\mathrm{DL}}^{(\mathrm{T})}$ of the L detector mode, and the time-independent matrix element for spontaneous photon emission into the mode labelled by $k^{(-)}$, μ is obtained in the form of (150) with

$$V_{fi}^{(T)}(\boldsymbol{K}^{(-)}, \mu) = -i\frac{e}{n}\left[\frac{\hbar}{2(2\pi)^3 K\epsilon_0}\right]^{1/2}$$

$$\times \left[\hat{\boldsymbol{\varepsilon}}(\hat{\boldsymbol{s}}^{(-)}, \mu) \cdot \boldsymbol{j}_{fi}(K\hat{\boldsymbol{s}}^{(-)})\right] T_{\mathrm{L}}(s_z, \mu) \exp(-iK\hat{\boldsymbol{s}}^{(-)} \cdot \boldsymbol{R}) . \quad (198)$$

It is noted that the atom interacts with homogeneous waves for $0 \le \kappa_\| < 1/n$ and with evanescent waves for $1/n \le \kappa_\| < 1$.

In order to evaluate the probability of multipole radiation, we introduce the multipole expansion of the transition current given by

$$\boldsymbol{j}_{fi}(K\hat{\boldsymbol{s}}^{(\pm)}) = \sum_{\lambda=1}^{3}\sideset{}{'}\sum_{j,m}\boldsymbol{Y}_{j,m}^{(\lambda)}(\hat{\boldsymbol{s}}^{(\pm)})J_{fi}(\lambda, K, j, -m) , \quad (199)$$

where $J_{fi}(\lambda, K, j, -m)$ are the expansion coefficients defined by

$$J_{fi}(\lambda, K, j, -m) = \int \boldsymbol{U}_{K,j,m}^{(\lambda)*}(\boldsymbol{r}_0) \cdot \boldsymbol{j}_{fi}(\boldsymbol{r}_0)\mathrm{d}^3x_0 , \quad (200)$$

with $\boldsymbol{Y}_{j,m}^{(\lambda)}(\hat{\boldsymbol{s}}^{(\pm)})$ and $\boldsymbol{U}_{K,j,m}^{(\lambda)*}(\boldsymbol{r})$, respectively, the momentum and position representations of vector spherical waves introduced in Appendix A. The multipole expansion of (199) can be obtained by using (251) in Appendix B. These vector mode functions are specified by parity λ, $\lambda = 1$ for Electric, $\lambda = 2$ for Magnetic, or $\lambda = 3$ for Longitudinal, wave number K, total angular momentum j, and its z projection, i.e., magnetic quantum number, m. The scalar product of the polarization vector and the Fourier component of the transition current can be rewritten as

$$\hat{\boldsymbol{\varepsilon}}(\hat{\boldsymbol{s}}^{(\pm)}, \mu) \cdot \boldsymbol{j}_{fi}(K\hat{\boldsymbol{s}}^{(\pm)}) =$$

$$\sum_{\lambda=1}^{2}\sum_{j}\sum_{m} f_{j,m}^{(\lambda)}(\hat{\boldsymbol{s}}^{(\pm)}, \mu)J_{fi}(\lambda, K, j, -m) , \quad (201)$$

with the expansion coeffcients defined as shown in Appendix B as

$$f_{j,m}^{(\lambda)}(\hat{\boldsymbol{s}}^{(\pm)}, \mu) = \hat{\boldsymbol{\varepsilon}}(\hat{\boldsymbol{s}}^{(\pm)}, \mu) \cdot \boldsymbol{Y}_{j,m}^{(\lambda)}(\hat{\boldsymbol{s}}^{(\pm)}) . \quad (202)$$

It is noted that $f_{j,m}^{(3)}(\hat{\boldsymbol{s}}^{(\pm)}, \mu) = 0$ for $\mu = 1$, 2, so that the photon-emission processes are irrelevant to the longitudinal component of $\lambda = 3$. According to (262) and (265) given in Appendix C, (200) can be represented as

$$J_{fi}(\lambda, K, j, -m) = (-1)^{j+m+\lambda}\,\mathrm{i}^{j+\lambda+1}$$

$$\times \sqrt{4\pi}\sqrt{\frac{(2j+1)(j+1)}{j}}\frac{K^j}{(2j+1)!!}Q_{fi}(\lambda, j, -m) , \quad (203)$$

where $\lambda = 1$, 2, and $Q_{fi}(\lambda, j, -m)$ indicates the 2^J-pole moments corresponding to the transition defined by (263) and (266) in Appendix C.

Now we consider the spontaneous-emission probability for the well-defined photonic final state with the angular momentum j, its z projection m, and parity λ, $\lambda = 1$ for Electric, $\lambda = 2$ for Magnetic. The replacement

$$\hat{\varepsilon}(\hat{s}^{(\pm)}, \mu) \cdot \boldsymbol{j}_{fi}(K\hat{s}^{(\pm)}) \rightarrow f_{j,m}^{(\lambda)}(\hat{s}^{(\pm)}, \mu) J_{fi}(\lambda, K, j, -m) \tag{204}$$

and substitution of (195) and (196) into (145) and (146) yield $d\Gamma_{+}^{(h)}$ and $d\Gamma_{+}^{(c)}$ as follows;

$$d\Gamma_{+}^{(h)}(\hat{s}^{(+)}, \mu) = \Gamma^{(0)}(\lambda, j, -m) \left\{ f_{j,m}^{(\lambda)*}(\hat{s}^{(+)}, \mu) f_{j,m}^{(\lambda)}(\hat{s}^{(+)}, \mu) \right.$$
$$\left. + f_{j,m}^{(\lambda)*}(\hat{s}^{(-)}, \mu) f_{j,m}^{(\lambda)}(\hat{s}^{(-)}, \mu) R_{\mathrm{R}}^2(s_z, \mu) \right\} d\Omega(\hat{s}^{(+)}) , \tag{205}$$

$$d\Gamma_{+}^{(c)}(\hat{s}^{(+)}, \mu) = 2\Gamma^{(0)}(\lambda, j, -m) \Re e \left\{ f_{j,m}^{(\lambda)*}(\hat{s}^{(+)}, \mu) f_{j,m}^{(\lambda)}(\hat{s}^{(-)}, \mu) \right.$$
$$\left. \times R_{\mathrm{R}}(s_z, \mu) \exp(2\mathrm{i}K s_z Z) \right\} d\Omega(\hat{s}^{(+)}) , \tag{206}$$

where $\rho = KZ$ is the normalized atom-to-boundary distance and

$$\Gamma^{(0)}(\lambda, j, -m)$$
$$= \frac{1}{4\pi\epsilon_0 \hbar} \frac{2(2j+1)(j+1)}{j\left[(2j+1)!!\right]^2} K^{2j+1} e^2 \left| Q_{fi}(\lambda, j, -m) \right|^2 . \tag{207}$$

Substituting (198) into (154) and (155), we obtain the differential transition probability corresponding, respectively, to the radiation of homogeneous photon, $d\Gamma_{-}^{(h)}$, and the photon tunneling via evanescent waves, $d\Gamma_{-}^{(t)}$, as follows;

$$d\Gamma_{-}^{(h)}(\hat{\kappa}^{(-)}, \mu) = \Gamma^{(0)}(\lambda, j, -m) f_{j,m}^{(\lambda)*}(\hat{s}^{(-)}, \mu) f_{j,m}^{(\lambda)}(\hat{s}^{(-)}, \mu)$$
$$\times nT_{\mathrm{L}}^2(s_z, \mu) d\Omega(\hat{\kappa}^{(-)}) , \tag{208}$$

$$d\Gamma_{-}^{(t)}(\hat{\kappa}^{(-)}, \mu) = \Gamma^{(0)}(\lambda, j, -m) f_{j,m}^{(\lambda)*}(\hat{s}^{(+)}, \mu) f_{j,m}^{(\lambda)}(\hat{s}^{(-)}, \mu)$$
$$\times nT_{\mathrm{L}}(-s_z, \mu) T_{\mathrm{L}}(s_z, \mu) \exp(2\mathrm{i}K s_z Z) d\Omega(\hat{\kappa}^{(-)}) . \tag{209}$$

6.2 Spontaneous Decay Rate of Multipoles in Half-Space

The probability of spontaneous emission into the right half-space associated with the atomic transition from the initial $|\varphi_i\rangle$ to final $|\varphi_f\rangle$ states is obtained by integrating $d\Gamma_+$ over the hemisphere $\mathcal{A}_+$, and summing up the contributions from all the polarizations μ. According to (158), the result is described as the sum of the radiation probability $\Gamma_{+}^{(h)}$ and its modulation $\Gamma_{+}^{(c)}$ due to interference as follows;

$$\Gamma_+^{(\mathrm{h})} = \Gamma^{(0)}(\lambda, j, -m) \sum_{\mu=1}^{2} \int\int_{0 \le s_\| < 1} \mathrm{d}s_x \mathrm{d}s_y \frac{1}{s_z}$$

$$\times \left[f_{j,m}^{(\lambda)*}(\hat{\boldsymbol{s}}^{(+)}, \mu) f_{j,m}^{(\lambda)}(\hat{\boldsymbol{s}}^{(+)}, \mu) \right.$$

$$\left. + f_{j,m}^{(\lambda)*}(\hat{\boldsymbol{s}}^{(-)}, \mu) f_{j,m}^{(\lambda)}(\hat{\boldsymbol{s}}^{(-)}, \mu) R_{\mathrm{R}}^2(s_z, \mu) \right] , \quad (210)$$

$$\Gamma_+^{(\mathrm{c})} = 2\Gamma^{(0)}(\lambda, j, -m) \Re e \left\{ \sum_{\mu=1}^{2} \int\int_{0 \le s_\| < 1} \mathrm{d}s_x \mathrm{d}s_y \frac{1}{s_z} \right.$$

$$\left. \times f_{j,m}^{(\lambda)*}(\hat{\boldsymbol{s}}^{(+)}, \mu) f_{j,m}^{(\lambda)}(\hat{\boldsymbol{s}}^{(-)}, \mu) R_{\mathrm{R}}(s_z, \mu) \exp(2\mathrm{i}K s_z Z) \right\} . \quad (211)$$

On the other hand, the probability of spontaneous emission into the left half-space is obtained by integrating $\mathrm{d}\Gamma_-$ over $\mathcal{A}_-$, and summing up all the polarization conponents μ. The result is described as the sum of two components according to (161); one is due to radiation of homogeneous waves $\Gamma_-^{(\mathrm{h})}$ and the other is due to radiation of evanescent waves interpreted as an optical excitation tunneling $\Gamma_-^{(\mathrm{t})}$ given, respectively, by

$$\Gamma_-^{(\mathrm{h})} = \Gamma^{(0)}(\lambda, j, -m) \sum_{\mu=1}^{2} \int\int_{0 \le s_\| < 1} \mathrm{d}s_x \mathrm{d}s_y \frac{1}{s_z}$$

$$\times f_{j,m}^{(\lambda)*}(\hat{\boldsymbol{s}}^{(-)}, \mu) f_{j,m}^{(\lambda)}(\hat{\boldsymbol{s}}^{(-)}, \mu) \left[1 - R_{\mathrm{R}}^2(s_z, \mu) \right] , \quad (212)$$

$$\Gamma_-^{(\mathrm{t})} = 2\Gamma^{(0)}(\lambda, j, -m) \Re e \left\{ \sum_{\mu=1}^{2} \int\int_{1 \le s_\| < n} \mathrm{d}s_x \mathrm{d}s_y \frac{1}{s_z} \right.$$

$$\left. \times f_{j,m}^{(\lambda)*}(\hat{\boldsymbol{s}}^{(+)}, \mu) f_{j,m}^{(\lambda)}(\hat{\boldsymbol{s}}^{(-)}, \mu) R_{\mathrm{R}}(s_z, \mu) \exp(2\mathrm{i}K s_z Z) \right\} . \quad (213)$$

The probability of spontaneous emission, i.e., the decay rate of excited multipoles, in half-space is enhanced compared with that in free space $\Gamma^{(0)}$ by $\Delta\Gamma$ given by (169). Substituting (210) and (212) into (168), and replacing variables of integration $\mathrm{d}s_x$ and $\mathrm{d}s_y$ by α and β under the transform $s_x = \sin\alpha\cos\beta$ and $s_y = \sin\alpha\sin\beta$, we obtain $\Gamma^{(0)}$ for multipole radiation as

$$\Gamma^{(0)} = \Gamma_+^{(\mathrm{h})} + \Gamma_-^{(\mathrm{h})} = \Gamma^{(0)}(\lambda, j, -m) , \quad (214)$$

where we have utilized the following narmalization condition of vector spherical harmonics;

$$\sum_{\mu=1}^{2} \int_0^\pi \sin\alpha\mathrm{d}\alpha \int_0^{2\pi} \mathrm{d}\beta f_{j,m}^{(\lambda)*}(\hat{\boldsymbol{s}}^{(+)}, \mu) f_{j,m}^{(\lambda)}(\hat{\boldsymbol{s}}^{(+)}, \mu) = 1 \quad \text{for } \lambda = 1, 2 .$$
$$(215)$$

Equation (214) corresponds to the spontaneous-emission rate $\Gamma^{(0)}$ in free space obtained in the limit of $n \to 1$ because $\Gamma_+^{(c)} = \Gamma_-^{(t)} = 0$ for $R_R(s_z, \mu) \to 0$. Equations (211) and (213) can be transformed, respectively, as

$$
\Gamma_+^{(c)} = 2\Gamma^{(0)}(\lambda, j, -m) \Re \left\{ \sum_{\mu=1}^{2} \int_0^{\pi/2} \sin \alpha d\alpha \int_0^{2\pi} d\beta \right.
$$
$$
\left. \times f_{j,m}^{(\lambda)*}(\hat{s}^{(+)}, \mu) f_{j,m}^{(\lambda)}(\hat{s}^{(-)}, \mu) R_R(s_z, \mu) \exp(2iK s_z Z) \right\} , \quad (216)
$$

$$
\Gamma_-^{(t)} = 2\Gamma^{(0)}(\lambda, j, -m) \Re \left\{ \sum_{\mu=1}^{2} \int_{\pi/2}^{(\pi/2)-i\gamma_c} \sin \alpha d\alpha \int_0^{2\pi} d\beta \right.
$$
$$
\left. \times f_{j,m}^{(\lambda)*}(\hat{s}^{(+)}, \mu) f_{j,m}^{(\lambda)}(\hat{s}^{(-)}, \mu) R_R(s_z, \mu) \exp(2iK s_z Z) \right\} . (217)
$$

According to (169), the optical near-field enhancement of the spontaneous decay rate of an atomic multipole $\Delta\Gamma$ is given by

$$
\Delta\Gamma = 2\Gamma^{(0)}(\lambda, j, -m) \Re \left\{ \sum_{\mu=1}^{2} \int_0^{(\pi/2)-i\gamma_c} \sin \alpha d\alpha \int_0^{2\pi} d\beta \right.
$$
$$
\left. \times f_{j,m}^{(\lambda)*}(\hat{s}^{(+)}, \mu) f_{j,m}^{(\lambda)}(\hat{s}^{(-)}, \mu) R_R(s_z, \mu) \exp(2iK s_z Z) \right\} . (218)
$$

As a numerical example, the spontaneous-emission probability or spontaneous decay rate of an electric quadrupole is shown in Fig. 22 for the case of the radiation corresponding to $\lambda = 1$ and $j = 2$, where the expansion coefficients are given as follows;

$$
f_{2,m}^{(1)}(\hat{s}^{(\pm)}, 1) = -i \left(\frac{5}{16\pi} \right)^{1/2}
$$
$$
\times \left[-\sin \alpha \exp(2i\beta)\delta_{m,+2} \pm \cos \alpha \exp(i\beta)\delta_{m,+1} \right.
$$
$$
\left. \pm \cos \alpha \exp(-i\beta)\delta_{m,-1} + \sin \alpha \exp(-2i\beta)\delta_{m,-2} \right] , \quad (219)
$$
$$
f_{2,m}^{(1)}(\hat{s}^{(\pm)}, 2) = - \left(\frac{5}{16\pi} \right)^{1/2}
$$
$$
\times \left[\mp \sin \alpha \cos \alpha \exp(2i\beta)\delta_{m,+2} - (1 - 2\cos \alpha^2) \exp(i\beta)\delta_{m,+1} \right.
$$
$$
\pm \sqrt{6} \sin \alpha \cos \alpha \delta_{m,0} + (1 - 2\cos^2 \alpha) \exp(-i\beta)\delta_{m,-1}
$$
$$
\left. \mp \sin \alpha \cos \alpha \exp(-2i\beta)\delta_{m,-2} \right] . \quad (220)
$$

As we discussed in the previous sections, the enhancement and modulation of spontaneous emission in half-space problems are due, respectively, to the interference of propagating waves in R-detector mode and the excitation transfer via evanescent waves in L-detector mode.

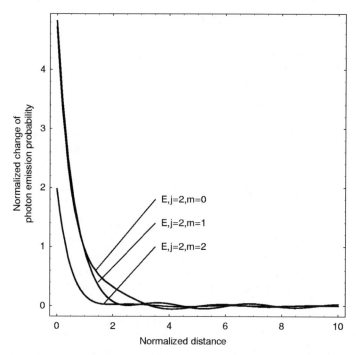

E,j=2,m=0

E,j=2,m=1

E,j=2,m=2

Fig. 22. Numerical results of normalized spontaneous-emission probability from the atomic quadrupole as a function of the normalized distance KZ. The orientation of the atomic quadrupole is indicated for each curve with respect to the quantization axis normal to the surface

As an example of a numerical result, we show in Fig. 22 the normalized spontaneous-emission probability calculated for the atomic quadrupole radiation near a planar dielectric surface with refractive index $n = 1.6$. As indicated in the figure, each curve corresponds to a different orientation of the atomic quadrupole with respect to the quantization axis normal to the surface. The interaction between the atomic quadrupole and the evanescent wave involved in the L-detector mode is strongly enhanced in the near-field regime, $KZ < 1$. The oscillation of the result observed for $KZ > 1$ is due to the interference between the direct and reflected radiation from the atomic quadrupole corresponding to the interaction with the homogeneous wave involved in the R-detector mode.

It is instructive to evaluate $\Delta\Gamma$ in the far-field limit, $(KZ \gg 1)$ corresponding to large atom-to-surface distances, where the integrand of (216) is rapidly oscillating due to the phase factor $2Ks_zZ$. Therefore, in this limit, we obtain $\Gamma_+^{(c)} \to 0$. Also, since the interaction via evanescent waves is weakened by $\exp(-2K|s_z|Z)$ in (217), $\Gamma_-^{(t)} \to 0$ in the far-field limit. In this way, the spontaneous decay rate of atomic multipoles in half-space problems approaches the value in free space in the far-field limit.

It should be noted that, according to the results obtained in the above, high-order multipoles exert stronger coupling with evanescent waves involved in L-detector mode during optical near-field interaction.

It is also noted that the theoretical treatments based on the detector modes developed in this section are also applicable to problems with spherical and cylindrical boundaries because of the use of angular-spectrum representation by which one can easily transform different representations utilizing the analytic nature of the angular spectrum or introduction of evanescent waves. Some examples of such transforms have been shown in the literature [29].

7   Tunneling Picture of Optical Near-Field Interactions

As we have discussed in the previous sections, the optical near-field interactions associated with energy-transport processes can be described as a tunneling of excitation via evanescent electromagnetic fields. This picture is especially useful in the investigation of the fundamental processes involved in optical near-field microscopy and nanometer-sized optoelectronic devices. In this section, we will introduce the tunneling picture based on the calculation of the Poynting vector of scattered fields using angular-spectrum representation of electromagnetic fields. Since the interacting objects are separated by an assumed planar boundary in the angular-spectrum representation, we can clearly evaluate the energy transfer between the objects in terms of the surface integration of the Poynting vector over the boundary plane. This implies that even in the optical near-field problems one can introduce a clear identification of optical source and sink, which provides us with the basis to investigate the signal transport and associated dissipation processes in the general nano-optics devices. It is stressed that the energy transfer of the tunneling regime takes place only through the overlap integral of evanescent waves with the same penetration depth and pseudomomentum involved in the angular spectra of scattered fields of interacting objects. In the following, we will clarify the role of dissipation processes that actually determine the transport of electromagnetic excitation. Generally speaking, it is the dissipation process, or explicitly the assumed absorbing boundary and source distribution, that determines the electromagnetic fields in numerical treatments of electromagnetic boundary problems, such as that based on the finite-difference time-domain algorithum.

Tunneling phenomena of the electron are usually described in terms of the tunneling current of Bardeen corresponding to the quantum-mechanical transition current composed of a product of the electronic wave functions defined in each half-space lying in the "right" and "left" of the potential barrier including the common interaction region where the potential barrier exists. The tunneling current provides the transition amplitude when it is integrated over an arbitrary surface lying in the middle of the barrier region. According to the golden rule of Fermi, the probabilitiy of excitation

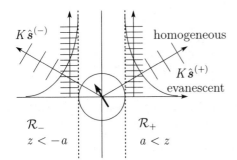

$K\hat{s}^{(-)}$ homogeneous

$K\hat{s}^{(+)}$
evanescent

$\mathcal{R}_-$ $\mathcal{R}_+$
$z < -a$ $a < z$

Fig. 23. Angular-spectrum representation of scattered fields

transfer is determined not only by the transition amplitudes, but also by the density of final states of the quantum-mechanical transition including both the electronic and photonic states. In the case of electron tunneling such as in scanning tunneling microscopes one actually considers a thermodynamical open-system where one assumes the existence of reserves of different chemical potentials corresponding to the bias across the tunnel barrier. In any case, the tunneling process is irreversible due only to dissipation processes.

Optical near-field interactions can also be viewed as a tunneling phenomenon when it is reduced to a one-dimensional problem, except for the statistics of tunneling particles, based on the angular-spectrum representation of scattered fields. In the case of the interactions between a pair of spatially localized subwavelength-sized objects placed in a subwavelength vicinity to each other, both the over-barrier transport of excitation via homogeneous waves and tunneling of excitation via evanescent waves are involved in the angular spectrum. In the tunneling regime the transport of electromagnetic energy takes place through the coupling of evanescent waves each of which is connected to the fields in the right or left half-space. In the optical case the energy transport through the barrier region can be evaluated in terms of the Poynting vector. Also in the case of optical excitations, the tunneling processes are determined by both the density of the final state and the dissipation process due to absorbers and photodetectors actually placed in the near-field region or those assumed implicitly in the far-field region.

7.1 Energy Transport via Tunneling in Optical Near-Field Interactions

To summarize the results obtained in the previous sections, the electromagnetic fields scattered by a small dielectric object involve both homogeneous and evanescent waves as described in the angular-spectrum representation given by

$$\boldsymbol{E}_{\mathrm{s}}(\boldsymbol{r}) = \left(\frac{\mathrm{i}K^3}{8\pi^2\varepsilon_0}\right) \sum_{\mu=1}^{2} \int\int_{-\infty}^{+\infty} \mathrm{d}s_x \mathrm{d}s_y \frac{1}{s_z}$$

$$\times \left[\varepsilon(\hat{\boldsymbol{s}}^{(\pm)},\mu) \cdot \boldsymbol{d}(K\hat{\boldsymbol{s}}^{(\pm)})\right] \varepsilon(\hat{\boldsymbol{s}}^{(\pm)},\mu) \exp\left(\mathrm{i}K\hat{\boldsymbol{s}}^{(\pm)} \cdot \boldsymbol{r}\right) \text{ in } \mathcal{R}_{\pm} . \quad (221)$$

Here, $\boldsymbol{d}(K\hat{\boldsymbol{s}}^{(\pm)})$ stands for the Fourier amplitude of the electric dipole induced in the scatterer by the total electric field of the system, $\boldsymbol{E}$;

$$\boldsymbol{d}(K\hat{\boldsymbol{s}}^{(\pm)}) = \varepsilon_0 \int \mathrm{d}^3 r' \exp(-\mathrm{i}K\hat{\boldsymbol{s}}^{(\pm)} \cdot \boldsymbol{r}')\chi(\boldsymbol{r}')\boldsymbol{E}(\boldsymbol{r}') , \quad (222)$$

where $\chi(\boldsymbol{r}')$ is the susceptibility of the dielectric object.

In general, the energy transport via optical near-field interactions of two scatterers is described in terms of the Poynting vector as

$$I^{(\mathrm{t})} = 2\varepsilon_0 \int\int_{-\infty}^{+\infty} \mathrm{d}x\mathrm{d}y\Re e\left\{\boldsymbol{E}_1(\boldsymbol{r}) \times \boldsymbol{B}_2^*(\boldsymbol{r}) + \boldsymbol{E}_2(\boldsymbol{r}) \times \boldsymbol{B}_1^*(\boldsymbol{r})\right\} \cdot \hat{\boldsymbol{e}}_z ,$$

$$(223)$$

where $\boldsymbol{E}_1$ and $\boldsymbol{E}_2$ indicate the complex amplitudes of the scattered waves from objects 1 and 2, respectively, and $\boldsymbol{B}_1$ and $\boldsymbol{B}_2$ are those of the associated magnetic fields. We have shown that $I^{(\mathrm{t})}$ corresponds to the energy transfer due to the coupling of two scatterers via evanescent waves. During energy transfer the selection rule holds for the wavevectors parallel to the boundary surface, which corresponds to the pseudomomentum conservation. This also implies that such a tunneling energy transfer via evanescent waves of a certain pseudomomentum is significant only when the distance between the scatterers comes within the penetration depth of the evanescent waves. Regarding the shape of the angular-spectrum of the scattered field, we can conclude that the tunneling energy transfer is significant when two scatterers of about the same size are placed at a distance similar to the sizes of the scatterers.

For the case of the interaction between a planar dielectric surface and a small scatterer as illustrated in Fig. 24, $\boldsymbol{E}_1$ corresponds to a transmitted wave produced directly from the incident wave with pseudomomentum $K\hat{\boldsymbol{s}}_{i\|}$ from the left side of the boundary, and $\boldsymbol{E}_2$ a scattered wave from the small object. We can describe $I^{(\mathrm{t})}$ as the product of the transmitted evanescent wave and the evanescent wave with pseudomomentum $K\hat{\boldsymbol{s}}_{i\|}$ involved in the angular spectrum of $\boldsymbol{E}_2$;

$$I^{(\mathrm{t})} = \Re e\Big\{(-\mathrm{i}K)T_{\mathrm{L}}(-s_{iz},\mu_i)$$

$$\times \left[\varepsilon(\hat{\boldsymbol{s}}_i^{(-)},\mu_i) \cdot \boldsymbol{d}_2(K\hat{\boldsymbol{s}}_i^{(-)})\right] \exp(-\mathrm{i}K\hat{\boldsymbol{s}}_i^{(-)} \cdot \boldsymbol{R})\Big\} . \quad (224)$$

The induced dipole $\boldsymbol{d}_2$ in the scatterer is expanded as

$$\boldsymbol{d}_2(K\hat{\boldsymbol{s}}_i^{(-)}) = \boldsymbol{d}_2^{(0)}(K\hat{\boldsymbol{s}}_i^{(-)}) + \boldsymbol{d}_2^{(1)}(K\hat{\boldsymbol{s}}_i^{(-)}) + \boldsymbol{d}_2^{(2)}(K\hat{\boldsymbol{s}}_i^{(-)}) + \cdots , \quad (225)$$

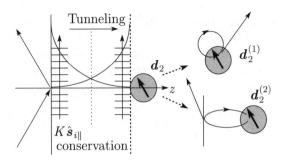

Fig. 24. Tunneling energy transfer between a planar and a small scatterer

where $\boldsymbol{d}_2^{(0)}$ is an induced dipole by the incident wave $\boldsymbol{E}_1$, which then induces two image dipoles in the small scatterer, $\boldsymbol{d}_2^{(1)}$ and $\boldsymbol{d}_2^{(2)}$. Substituting this into (224) and considering each component, we can find the fundamental processes involved in the interaction illustrated in Fig. 24. The component of $I^{(\mathrm{t})}$ resulting from $\boldsymbol{d}_2^{(0)}$ yields zero energy transfer. This can be understood from the fact that the equiphase planes of evanescent waves are perpendicular to the boundary plane, so that the reflection of the evanescent wave directly to the induced dipole in the scatterer makes no difference to the total reflection at the dielectric surface. Nonzero energy transfer results from the components of $I^{(\mathrm{t})}$ due to $\boldsymbol{d}_2^{(1)}$ and $\boldsymbol{d}_2^{(2)}$. For each component the corresponding interaction process is schematically indicated in Fig. 24. The self-interaction resulting in $\boldsymbol{d}_2^{(1)}$ corresponds to the reorganization of induced electric dipole to produce the self-consistent distributions of the fields and polarizations in the scatterer of finite spatial extension. From the viewpoint of the angular-spectrum representation, the small object scatters the incident evanescent wave into an evanescent wave with different pseudomomentum, which, in turn, is scattered by the small object into the propagating wave into the far-field region. In each process of scattering, the small object receives the recoil momentum from the scattered field according to the restricted conservation law of momentum in the direction parallel to the dielectric surface. $\boldsymbol{d}_2^{(2)}$ is due to the higher-order interaction between the scatterer and the dielectric surface. Also in this case, the scattering process is due to evanescent waves of different pseudomomenta involved in the angular spectrum of the scattered field of the small object. We have shown that the scattered field due to $\boldsymbol{d}_2^{(1)}$ contributes to the modulation of the reflected fields via interference with the reflected incident field $\boldsymbol{d}_2^{(0)}$. The scattered field due to $\boldsymbol{d}_2^{(2)}$ contributes to the intensity reflected back into the medium via the tunneling energy transfer. This process involves the propagating wave with the same propagation direction as the totally reflected incident wave at the boundary, which decreases the intensity of the reflected light due to interference so as to maintain the energy conservation. In other words, the contribution from $\boldsymbol{d}_2^{(2)}$ is the counterpart of the radiation from the small object. It is stressed that the nonzero tunneling energy transfer is due to these image dipoles.

For the case of the interaction between two small scatterers, we consider $\boldsymbol{E}_1$ and $\boldsymbol{E}_2$ as the complex amplitude corresponding, respectively, to the scattered waves from the scatterers 1 and 2. The tunneling energy transfer $I^{(\mathrm{t})}$ is given by the coupling of evanescent waves with the same pseudomomentum in the angular spectra of $\boldsymbol{E}_1$ and $\boldsymbol{E}_2$;

$$
I^{(\mathrm{t})} = \left(\frac{K^4}{4\pi^2\varepsilon_0} \right) \Re e \left\{ \sum_{\mu=1}^{2} \int\!\!\!\int_{1\leq s_\parallel} \frac{\mathrm{d}s_x \mathrm{d}s_y}{s_z} \left[\boldsymbol{\varepsilon}(\hat{\boldsymbol{s}}^{(+)}, \mu) \cdot \boldsymbol{d}_1(K\hat{\boldsymbol{s}}^{(+)}) \right]^* \right.
$$
$$
\left. \times \left[\boldsymbol{\varepsilon}(\hat{\boldsymbol{s}}^{(-)}, \mu) \cdot \boldsymbol{d}_2(K\hat{\boldsymbol{s}}^{(-)}) \right] \exp\left[\mathrm{i}K\hat{\boldsymbol{s}}^{(-)} \cdot (\boldsymbol{R}_1 - \boldsymbol{R}_2) \right] \right\} . \quad (226)
$$

Only when some modulations of $\boldsymbol{d}_1$ are induced on $\boldsymbol{d}_1^{(0)}$ by an additional incoming wave due to near-field interactions with the environment in addition to some dissipation processes, does the tunneling energy have a nonzero value, which is actually due to the image dipoles $\boldsymbol{d}_2^{(1)}$ and $\boldsymbol{d}_2^{(2)}$ involved in $\boldsymbol{d}_2$.

7.2   Fundamental Process in Nano-Optics Device

The fundamental processes of the optical near-field interaction described by (226) is illustrated in Fig. 25, which corresponds to a nanophotonic device exerting its function via optical near-field excitation transfer between elements. The tunneling picture developed in this chapter provides general bases for investigations of such devices with a clear understanding of the important role of dissipation in the process of the directional signal transport properties. As an example, we show in Fig. 26 a schematic diagram of a scanning optical near-field microscope, where the fundamental processes involved are described in a similar way to those in Fig. 25 by the scattering and interactions between four major objects, i.e., a planar dielectric substrate, sample object, probe tip, and tapered waveguide. It is again stressed that, even in this case, the excitation transfer is determined by the dissipation process at the sink and by transmission loss due to stray propagating fields scattered into far field where sinks are assumed implicitly. The theoretical study in this chapter provides us with sufficient description of the fundamental processes to evaluate the entire process of near-field optical microscopes shown in Fig. 26, as well as general nano-optics devices.

An extension of discussion of the multipole excitation is especially important since the near-field interactions based on multipole transitions are less dissipative with respect to the radiation in the far-field region, so that one can find the meaning of local interaction via evanescent waves with higher wave numbers and short penetration depths even in a narrow space of nanometer size. Such a study provides the indispensable basis for investigation of the functions of nanometer-sized electronic or optoelectronic devices in terms of the signal transport and associated dissipation processes.

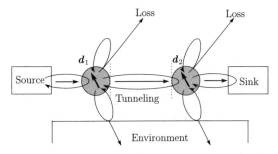

Fig. 25. Model of nanometer-sized optoelectronic device with the fundamental processes of optical near-field interactions viewed as a tunneling energy transfer as well as the dissipation processes that actually dominate the energy transport

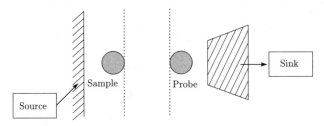

Fig. 26. Fundamental processes of optical near-field microscopy viewed as a nano-optoelectronic device: One can investigate the near-field interactions and dissipation processes on the basis of the model device shown in Fig. 25

Appendices

A   Vector Spherical Wave

The state of vector spherical waves is specified by using the following set of parameters; parity λ with $\lambda = 1$ for Electric, $\lambda = 2$ for Magnetic, $\lambda = 3$ for Longitudinal, wave number K, total angular momentum j, and its z projection m, i.e., magnetic quantum number [31–34]. The vector spherical waves are defined in the momentum representation as

$$U^{(\lambda)}_{K,j,m}(\boldsymbol{K}') = \frac{(2\pi)^3}{K^2}\delta(K' - K)\boldsymbol{Y}^{(\lambda)}_{j,m}(\hat{\boldsymbol{s}}') \,, \tag{227}$$

where the unit wavevector is $\hat{\boldsymbol{s}}' = (\boldsymbol{K}'/K)$ and the vector spherical harmonics are given by

$$\boldsymbol{Y}^{(1)}_{j,m}(\hat{\boldsymbol{s}}) = \frac{K}{\sqrt{j(j+1)}}\nabla Y^m_j(\hat{\boldsymbol{s}}) \,, \tag{228}$$

$$\boldsymbol{Y}^{(2)}_{j,m}(\hat{\boldsymbol{s}}) = -\mathrm{i}K\frac{\hat{\boldsymbol{s}} \times \nabla}{\sqrt{j(j+1)}}Y^m_j(\hat{\boldsymbol{s}}) \,, \tag{229}$$

$$\boldsymbol{Y}^{(3)}_{j,m}(\hat{\boldsymbol{s}}) = \hat{\boldsymbol{s}}Y^m_j(\hat{\boldsymbol{s}}) \,, \tag{230}$$

with the spherical harmonic Y_j^m and $\nabla = \partial/(\partial \boldsymbol{K})$ [35]. These orthogonality relations are described as

$$\int_0^\pi \sin\alpha\,d\alpha \int_0^{2\pi} d\beta\, \boldsymbol{Y}_{j,m}^{(\lambda)*}(\hat{\boldsymbol{s}}) \cdot \boldsymbol{Y}_{j',m'}^{(\lambda')}(\hat{\boldsymbol{s}}) = \delta_{\lambda\lambda'}\delta_{jj'}\delta_{mm'} . \tag{231}$$

In order to describe the states of polarization, it is convenient to introduce the other representation of vector spherical harmonics defined by

$$\boldsymbol{Y}_{j,m}^{(1)}(\hat{\boldsymbol{s}}) = \sqrt{\frac{j}{2j+1}}\,\boldsymbol{Y}_{j,j+1,m}(\hat{\boldsymbol{s}}) + \sqrt{\frac{j+1}{2j+1}}\,\boldsymbol{Y}_{j,j-1,m}(\hat{\boldsymbol{s}}) , \tag{232}$$

$$\boldsymbol{Y}_{j,m}^{(2)}(\hat{\boldsymbol{s}}) = \boldsymbol{Y}_{j,j,m}(\hat{\boldsymbol{s}}) , \tag{233}$$

$$\boldsymbol{Y}_{j,m}^{(3)}(\hat{\boldsymbol{s}}) = -\sqrt{\frac{j+1}{2j+1}}\,\boldsymbol{Y}_{j,j+1,m}(\hat{\boldsymbol{s}}) + \sqrt{\frac{j}{2j+1}}\,\boldsymbol{Y}_{j,j-1,m}(\hat{\boldsymbol{s}}) , \tag{234}$$

where

$$\boldsymbol{Y}_{j,\ell,m}(\hat{\boldsymbol{s}}) = \sum_{\mu=-1}^{+1} \langle \ell, 1; m-\mu, \mu|\, \ell, 1; j, m \rangle\, Y_\ell^{m-\mu}(\hat{\boldsymbol{s}})\hat{\boldsymbol{e}}_\mu , \tag{235}$$

with Clebsch–Gordan coefficients $\langle \ell, 1; m-\mu, \mu|\, \ell, 1; j, m \rangle$ describing the vector spherical harmonics specified by j and m as the composite state of the electric multipole with ℓ and the vector nature of electromagnetic fields with spin 1.

The states of a photon in the spatial representation are obtained by

$$\boldsymbol{U}_{K,j,m}^{(\lambda)}(\boldsymbol{r}) = \frac{1}{(2\pi)^3}\int d^3K'\,\exp(iK'\hat{\boldsymbol{s}}' \cdot \boldsymbol{r})\boldsymbol{U}_{K,j,m}^{(\lambda)}(\boldsymbol{K}') . \tag{236}$$

It is useful to introduce the following spherical wave representation of plane waves

$$\exp(iK\hat{\boldsymbol{s}} \cdot \boldsymbol{r}) = 4\pi \sum_{\ell=0}^{+\infty} \sum_{m=-\ell}^{+\ell} i^\ell j_\ell(Kr) Y_\ell^m(\hat{\boldsymbol{r}}) Y_\ell^{m*}(\hat{\boldsymbol{s}}) , \tag{237}$$

with unit vector $\hat{\boldsymbol{r}} = (\boldsymbol{r}/r)$ and spherical Bessel function $j_\ell(Kr)$. Substituting (227) and (237) into (236) and using the orthogonality relation of spherical harmonics, one can obtain

$$\boldsymbol{U}_{K,j,m}^{(1)}(\boldsymbol{r}) = 4\pi \left[i^{j+1}\sqrt{\frac{j}{2j+1}}\, j_{j+1}(Kr)\boldsymbol{Y}_{j,j+1,m}(\hat{\boldsymbol{r}}) \right.$$

$$\left. +i^{j-1}\sqrt{\frac{j+1}{2j+1}}\, j_{j-1}(Kr)\boldsymbol{Y}_{j,j-1,m}(\hat{\boldsymbol{r}}) \right] , \tag{238}$$

$$\boldsymbol{U}^{(2)}_{K,j,m}(\boldsymbol{r}) = 4\pi \mathrm{i}^j j_j(Kr)\boldsymbol{Y}_{j,j,m}(\hat{\boldsymbol{r}}) , \tag{239}$$

$$\boldsymbol{U}^{(3)}_{K,j,m}(\boldsymbol{r}) = 4\pi \left[-\mathrm{i}^{j+1}\sqrt{\frac{j+1}{2j+1}} j_{j+1}(Kr)\boldsymbol{Y}_{j,j+1,m}(\hat{\boldsymbol{r}}) \right.$$

$$\left. +\mathrm{i}^{j-1}\sqrt{\frac{j}{2j+1}} j_{j-1}(Kr)\boldsymbol{Y}_{j,j-1,m}(\hat{\boldsymbol{r}}) \right] . \tag{240}$$

The orthogonality relation between photonic states in position representation is given by

$$\int \mathrm{d}^3 x \boldsymbol{U}^{(\lambda)*}_{K,j,m}(\boldsymbol{r}) \cdot \boldsymbol{U}^{(\lambda')}_{K',j',m'}(\boldsymbol{r}) = \frac{(2\pi)^3}{K^2}\delta(K-K')\delta_{\lambda\lambda'}\delta_{jj'}\delta_{mm'} , \tag{241}$$

for which the following relations hold;

$$\boldsymbol{U}^{(\lambda)}_{K,j,m}(-\boldsymbol{r}) = (-1)^{j+\lambda}\boldsymbol{U}^{(\lambda)}_{K,j,m}(\boldsymbol{r}) , \tag{242}$$

$$\boldsymbol{U}^{(\lambda)*}_{K,j,m}(\boldsymbol{r}) = (-1)^{j+m+1}\boldsymbol{U}^{(\lambda)}_{K,j,-m}(\boldsymbol{r}) . \tag{243}$$

It is also useful to introduce scalar spherical waves defined by

$$\Phi_{K,j,m}(\boldsymbol{r}) = 4\pi \mathrm{i}^j j^j(Kr)Y_j^m(\hat{\boldsymbol{r}}) , \tag{244}$$

where the scalar spherical waves satisfy the following relation;

$$\mathrm{i}K\boldsymbol{U}^{(3)}_{K,j,m}(\boldsymbol{r}) = \nabla\Phi_{K,j,m}(\boldsymbol{r}) , \tag{245}$$

$$\Phi_{K,j,m}(-\boldsymbol{r}) = (-1)^j \Phi_{K,j,m}(\boldsymbol{r}) , \tag{246}$$

$$\Phi^*_{K,j,m}(\boldsymbol{r}) = (-1)^{j+m}\Phi_{K,j,-m}(\boldsymbol{r}) . \tag{247}$$

B   Expansion of the Vector Plane Wave in Terms of the Vector Spherical Waves

Scalar plane waves are described in terms of the vector spherical waves given by

$$\exp(\mathrm{i}K\hat{\boldsymbol{s}}\cdot\boldsymbol{r}) = \sum_{\lambda=1}^{3}\sum_{j,m}{}' \boldsymbol{U}^{(\lambda)}_{K,j,m}(\boldsymbol{r}) \otimes \boldsymbol{Y}^{(\lambda)*}_{j,m}(\hat{\boldsymbol{s}}) , \tag{248}$$

where the sum is taken as

$$\sum_{j,m}{}' = \begin{cases} \displaystyle\sum_{j=1}^{\infty}\sum_{m=-j}^{+j} & \text{for } \lambda = 1, 2 , \\[4mm] \displaystyle\sum_{j=0}^{\infty}\sum_{m=-j}^{+j} & \text{for } \lambda = 3 . \end{cases} \tag{249}$$

Replacing $\hat{s}$ in (248) by $\hat{s}^{(\pm)}$, we can obtain the expansion of vector evanescent waves. The vector spherical harmonics with the argument $\hat{s}^{(\pm)}$ can be obtained from the representation of (232)–(234). Replacing the unit wavevector $\hat{s}$ in (248) by $\hat{s}^{(\mp)*}$, we obtain

$$\exp(\mathrm{i}K\hat{s}^{(\mp)*}\cdot\boldsymbol{r}) = \sum_{\lambda=1}^{3}{\sum_{j,m}}'\boldsymbol{U}_{K,j,m}^{(\lambda)}(\boldsymbol{r})\otimes\boldsymbol{Y}_{j,m}^{(\lambda)*}(\hat{s}^{(\mp)*})\,, \tag{250}$$

where the unit wavevector takes $\hat{s}^{(-)*}$ for $z \geq 0$, and $\hat{s}^{(+)*}$ for $z < 0$. The complex conjugate of (250) is given by

$$\exp(-\mathrm{i}K\hat{s}^{(\mp)}\cdot\boldsymbol{r}) = \sum_{\lambda=1}^{3}{\sum_{j,m}}'\boldsymbol{Y}_{j,m}^{(\lambda)}(\hat{s}^{(\mp)})\otimes\boldsymbol{U}_{K,j,m}^{(\lambda)*}(\boldsymbol{r})\,, \tag{251}$$

which leads to the expansion of vector plane waves in terms of vector spherical waves as

$$\hat{\varepsilon}(\hat{s}^{(\mp)},\mu)\exp(-\mathrm{i}K\hat{s}^{(\mp)}\cdot\boldsymbol{r}) = \sum_{\lambda=1}^{2}\sum_{j,m}f_{j,m}^{(\lambda)}(\hat{s}^{(\mp)},\mu)\boldsymbol{U}_{K,j,m}^{(\lambda)*}(\boldsymbol{r})\,, \tag{252}$$

with expansion coefficients defined by

$$f_{j,m}^{(\lambda)}(\hat{s}^{(\mp)},\mu) = \hat{\varepsilon}(\hat{s}^{(\mp)},\mu)\cdot\boldsymbol{Y}_{j,m}^{(\lambda)}(\hat{s}^{(\mp)})\,. \tag{253}$$

Here we have used $\hat{\varepsilon}(\hat{s}^{(\mp)},\mu)\cdot\boldsymbol{Y}_{j,m}^{(3)}(\hat{s}^{(\mp)}) = 0$.

C Multipole Expansion of Transition Current

We consider multipole expansion of transition current (given in (200));

$$J_{fi}(\lambda,K,j,-m) = \int \mathrm{d}^3x\,\boldsymbol{U}_{K,j,m}^{(\lambda)*}(\boldsymbol{r})\cdot\boldsymbol{j}_{fi}(\boldsymbol{r})\,. \tag{254}$$

Using the relation given in (243), we obtain

$$J_{fi}(\lambda,K,j,-m) = (-1)^{j+m+1}\int \mathrm{d}^3x\,\boldsymbol{U}_{K,j,-m}^{(\lambda)}(\boldsymbol{r})\cdot\boldsymbol{j}_{fi}(\boldsymbol{r})\,. \tag{255}$$

Electric Multipole. We will first evaluate (255) for $\lambda = 1$ corresponding to electric multipole transitions. According to (242) the vector spherical waves with $\lambda = 1, 3$ are the same in parity, so that $J_{fi}(1,K,j,-m)$ can be transformed by using (245) as

$$J_{fi}(1,K,j,-m) = (-1)^{j+m+1}\int \mathrm{d}^3x\bigg\{\boldsymbol{U}_{K,j,-m}^{(1)}(\boldsymbol{r})$$

$$+ C\left[\boldsymbol{U}_{K,j,-m}^{(3)}(\boldsymbol{r}) - \frac{1}{\mathrm{i}K}\nabla\Phi_{K,j,-m}(\boldsymbol{r})\right]\bigg\}\cdot\boldsymbol{j}_{fi}(\boldsymbol{r})\,, \tag{256}$$

where C is an arbitrary constant. If we assume the transition current $\boldsymbol{j}_{fi}$ satisfies the continuity equation as

$$\nabla \cdot \boldsymbol{j}_{fi}(\boldsymbol{r}) = \mathrm{i}K\rho_{fi}(\boldsymbol{r}) , \qquad (257)$$

we can refer to $\rho_{fi}(\boldsymbol{r})$ as the transition density. Using the relation of (257) and considering that $\nabla(\varPhi\rho_{fi})$ vanishes as integrated over d^3x, (256) is rewritten as

$$J_{fi}(1, K, j, -m) = (-1)^{j+m+1} \int \mathrm{d}^3x \Big\{ C\varPhi_{K,j,-m}(\boldsymbol{r})\rho_{fi}(\boldsymbol{r})$$

$$+ \Big[\boldsymbol{U}^{(1)}_{K,j,-m}(\boldsymbol{r}) + C\boldsymbol{U}^{(3)}_{K,j,-m}(\boldsymbol{r}) \Big] \cdot \boldsymbol{j}_{fi}(\boldsymbol{r}) \Big\} , \qquad (258)$$

which corresponds to a gauge transform in electromagnetic interactions. When C is chosen as

$$C = -\sqrt{\frac{j+1}{j}} , \qquad (259)$$

the spherical Bessel functions of order $j - 1$ involved in $\boldsymbol{U}^{(1)}_{K,j,-m}(\boldsymbol{r})$ and $\boldsymbol{U}^{(3)}_{K,j,-m}(\boldsymbol{r})$ cancel each other out, so that there remains only the spherical Bessel function of order $j+1$ that contributes to $J_{fi}(1, K, j, -m)$ in the higher order with respect to Ka than the scalar function $\varPhi$ consisted of the spherical Bessel function of order j. As a result, in the near-field regime $Ka \ll 1$, we obtain

$$J_{fi}(1, K, j, -m) = (-1)^{j+m} \sqrt{\frac{j+1}{j}} \int \mathrm{d}^3x \varPhi_{K,j,-m}(\boldsymbol{r})\rho_{fi}(\boldsymbol{r}) , \qquad (260)$$

where only the contributions from $Kr \ll 1$ are significant in the integration with respect to d^3x. Raplacing $j_j(Kr)$ by its lowest-order term of Kr as

$$j_j(Kr) \sim \frac{(Kr)^j}{(2j+1)!!} , \qquad (261)$$

we obtain the transition current in the near-field regime as

$$J_{fi}(1, K, j, -m) =$$

$$(-1)^{j+m} \mathrm{i}^j 2\sqrt{\pi} \sqrt{\frac{(2j+1)(j+1)}{j}} \frac{K^j}{(2j+1)!!} Q_{fi}(1, j, -m) , \qquad (262)$$

where $Q_{fi}(1, j, -m)$ stands for the electric multipole transition moments defined by

$$Q_{fi}(1, j, -m) = \sqrt{\frac{4\pi}{2j+1}} \int \mathrm{d}^3x r^j Y_j^{-m}(\hat{\boldsymbol{r}})\rho_{fi}(\boldsymbol{r}) . \qquad (263)$$

Magnetic Multipole. We evaluate (255) for $\lambda = 2$ corresponding to magnetic multipole transitions. Substituting (229) into (239) with replacement of $\hat{s}$ by $\hat{r}$, the vector spherical wave for $\lambda = 2$ is obtained as

$$\boldsymbol{U}^{(2)}_{K,j,m}(\boldsymbol{r}) = -4\pi \mathrm{i}^{j+1} j_j(Kr) \frac{\boldsymbol{r} \times \nabla}{\sqrt{j(j+1)}} Y_j^m(\hat{\boldsymbol{r}}) \,, \tag{264}$$

where $\boldsymbol{r} = r\hat{\boldsymbol{r}}$. Substituting this into (255) for $\lambda = 2$ and adopting (261) in the near-field regime, we obtain the transition current for magnetic multipole transitions as

$$J_{fi}(2, K, j, -m) = (-1)^{j+m+1} \mathrm{i}^{j+1}$$

$$\times 2\sqrt{\pi} \sqrt{\frac{(2j+1)(j+1)}{j}} \frac{K^j}{(2j+1)!!} Q_{fi}(2, j, -m) \,, \tag{265}$$

where we have introduced the magnetic multipole transition moments given by

$$Q_{fi}(2, j, -m) = \frac{1}{(j+1)} \sqrt{\frac{4\pi}{2j+1}} \int \mathrm{d}^3 x \, [\boldsymbol{r} \times \boldsymbol{j}_{fi}(\boldsymbol{r})] \cdot \nabla \left(r^j Y_j^{-m}(\hat{\boldsymbol{r}}) \right) \,. \tag{266}$$

References

1. S. Kawata, M. Ohtsu, M. Irie (Eds): *Nano Optics* (Springer-Verlag, Berlin 2002)
2. E. Wolf, M. Nieto-Vesperinas: J. Opt. Soc. Am. A **2**, 886 (1985)
3. J.M. Vigoureux, R. Payen: J. Phys. (France) **36**, 1327 (1975)
4. H. Hori: In *Optical and Electronic Process of Nano-Matters* ed. by M. Ohtsu (KTK Sci. Publ., Tokyo/Kluwer Acad. Publ., Dordrecht 2001) pp. 1–55
5. M. Born, E. Wolf: *Principles of Optics* (Pergamon Press, 1975)
6. T. Inoue, H. Hori: Opt. Rev. **3**, 458 (1996)
7. M. Ohtsu, H. Hori: *Near-Field Nano-Optics* (Academic/Plenum, New York 1999) p. 300
8. P. Berman (Ed): *Cavity Quantum Electrodynamics* (Academic Press, San Diego 1994)
9. C.J. Hood, M.S. Chapman, T.W. Lynn, H.J. Kimble: Phys. Rev. Lett. **80**, 4157 (1998)
10. G.S. Agarwal: Phys. Rev. A **11**, 230 (1975); *ibid*, 243 (1975); *ibid*, 253 (1975)
11. J.M. Wylie, J.E. Sipe: Phys. Rev. A **30**, 1185 (1984)
12. S.D. Gupta, G.S. Agarwal: Opt. Commun. **115**, 597 (1995)
13. H. Chew: Phys. Rev. A **38**, 3410 (1988)
14. J. Ye, D.W. Vernoy, H.J. Kimble: Phys. Rev. Lett. **83**, 4987 (1999)
15. D.W. Vernooy, A. Furusawa, N. Ph Georgiades, V.S. Ilchenko, H.J. Kimble: Phys. Rev. A **57**, R2293 (1998)
16. C.K. Carniglia, L. Mandel, K.H. Drexhage: J. Opt. Soc. Am. **62**, 479 (1972)
17. T. Matsudo, T. Inoue, Y. Inoue, H. Hori, T. Sakurai: Phys. Rev. A **55**, 2406 (1997)

18. T. Matsudo, Y. Takahara, H. Hori, T. Sakurai: Opt. Commun. **145**, 64 (1998)
19. C.G. Aminoff, A.M. Steane, P. Bouyer, P. Desbiolles, J. Dalibard, C. Cohen-Tannoudji: Phys. Rev. Lett. **71**, 3083 (1993)
20. H. Ito, T. Nakata, K. Sakaki, M. Otsu, K.I. Lee, W. Jhe: Phys. Rev. Lett. **76**, 4500 (1996)
21. V.I. Balykin, V.S. Letokhov, Yu.B. Ovchinnikov, A.I. Sidorov: Phys. Rev. Lett. **60**, 2137 (1988)
22. C.K. Carniglia, L. Mandel: Phys. Rev. D **3**, 280 (1971)
23. W. Lukosz, R.E. Kunz: J. Opt. Soc. Am. **67** 1607 (1977)
24. W. Lukosz, R.E. Kunz: J. Opt. Soc. Am. **67**, 1615 (1977)
25. W. Lukosz: J. Opt. Soc. Am. **69**, 1495 (1979)
26. T. Inoue, H. Hori: Phys. Rev. A **63**, 063805 (2001)
27. W. Jhe, K. Jang: Phys. Rev. A **53**, 1126 (1996)
28. V.V. Klimov, V.S. Letokhov: Opt. Commun. **122**, 155 (1996)
29. T. Inoue, I. Banno, H. Hori: Opt. Rev. **5**, 295 (1998)
30. L.D. Landau, E.M. Lifshitz: *Quantum Mechanics (Non-relativistic Theory)*, 3rd edn (Pergamon Press, New York 1977)
31. J.D. Jackson: *Classical Electrodynamics*, 2nd edn (Wiley, New York 1975)
32. A.S. Davydov: *Quantum Mechanics*, 1st edn (Pergamon Press, Oxford 1965)
33. M.E. Rose: *Elementary Theory of Angular Momentum* (Wiley, New York 1957)
34. A.R. Edomonds: *Angular Momentum in Quantum Mechanics*, 2nd edn (Princeton University Press, Princeton 1974)
35. J.J. Sakurai: *Modern Quantum Mechanics*, 2nd edn (Benjamin/Cummings Publishing Company Inc., Reading, Mass. 1985)

Index

Springer Series in
OPTICAL SCIENCES

Springer Series in
OPTICAL SCIENCES

Printing: Krips bv, Meppel
Binding: Litges & Dopf, Heppenheim